Discrete Event Systems in Dioid Algebra and Conventional Algebra

FOCUS SERIES IN AUTOMATION & CONTROL

Series Editor Bernard Dubuisson

Discrete Event Systems in Dioid Algebra and Conventional Algebra

Philippe Declerck

iSTE

WILEY

First published 2013 in Great Britain and the United States by ISTE Ltd and John Wiley & Sons, Inc.

ISTE Ltd
27-37 St George's Road
London SW19 4EU
UK

www.iste.co.uk

John Wiley & Sons, Inc.
111 River Street
Hoboken, NJ 07030
USA

www.wiley.com

Library of Congress Control Number: 2012948947

British Library Cataloguing-in-Publication Data
A CIP record for this book is available from the British Library
ISSN: 2051-2481 (Print)
ISSN: 2051-249X (Online)
ISBN: 978-1-84821-461-3

MIX
Paper from responsible sources
FSC® C013604
www.fsc.org

Printed and bound in Great Britain by CPI Group (UK) Ltd., Croydon, Surrey CR0 4YY

Contents

1

Introduction

1.1. General introduction

The main motivation of this book is the modeling of a class of dynamic systems usually called "discrete event systems" where the timing of the events is crucial. Events are viewed as sudden changes in the process, which is essentially a man-made system, such as automated manufacturing lines or transportation systems. Forming the core of the following sections, the synchronization of the tasks is expressed by the maximum of the ends of the tasks conditioning the beginning of another task. A classical example is the departure time of a train that should wait for the arrival of other trains in order to allow for the changeover of passengers. The basic idea of dioid algebra is to treat the maximum as an operation that is included in algebra, here the (max, +) algebra. A large part of this book concerns this topic. Another fundamental idea is to treat the maximum in a maximization of a criterion as in linear programming. We also focus on this second topic in this book.

1.2. History and three mainstays

This book is based on the three following mainstays:

– Well developed in the European and French scientific community, the topic of dioid algebra as (max, +) algebra has existed for the past three decades. Its main advantage is its *formalism* that allows us to clearly describe complex notions and the possibility of transposing theoretical results between dioids and practical applications. Only a few books have been written on this subject. The first was written by R.A. Cuninghame-Green in 1979. We can quote in graph theory the important books of Gondran and Minoux (Chapter 3 in [GON 84] and [GON 02]). At the origin of the "French school", another book [BAC 92] was written from the system-theoretical point of view in 1992. Finally, the last book focuses on the transportation systems and was produced in 2006 by a team working in the Netherlands [HEI 06].

– In conventional algebra, linear programming is the oldest for which we can quote the famous French mathematician and physicist Joseph Fourier (1768–1830) for his work at the beginning of the 19th Century (Fourier–Motzkin algorithm). However, this topic is still current and the IT community has produced many algorithms that have efficiently solved bivariable linear systems for the past three decades.

– The third mainstay is the Petri net, which has its origin in Carl Adam Petri's dissertation submitted in 1962 in West Germany (Carl Adam Petri 1926–2010). A Petri net is a graphical modeling tool that describes complex discrete event systems with various themes and application fields. This topic has been well developed for the past three decades in Europe.

1.3. Scientific context

We describe below the main concepts that will determine the following studies in the mathematical field and Petri nets. The modeling of the time behaviors will also play an important role.

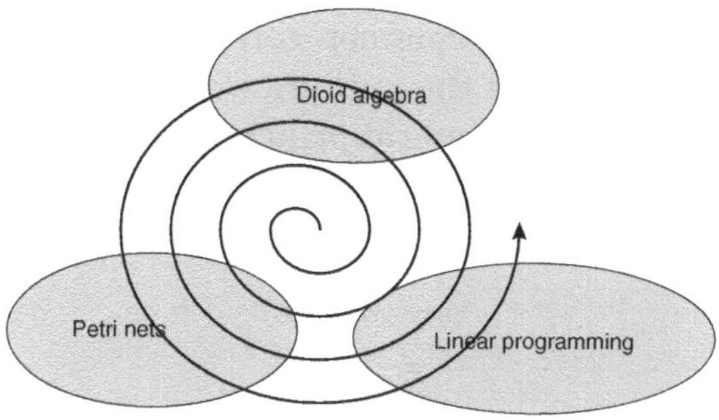

Figure 1.1. *Three topics*

1.3.1. *Dioids*

In the mathematical field, there are different structures with sets provided with one or two operations [GON 84]. We principally consider in this book the structure of the dioid, which is an idempotent semi-ring, and, in particular, the (max, +) algebra $(\mathbb{R} \cup \{-\infty\}, \oplus, \otimes)$, which is an example of a dioid. This is an interesting tool allowing us to study graph theory and to analyze discrete event systems. The structure of the ring is well-known as familiar examples are integers and real numbers, equipped with the usual addition and multiplication ($(\mathbb{Z}, +, \times)$ and $(\mathbb{R}, +, \times)$, respectively), which are used in many fields such as automatic control. The common point is the structure of the semi-ring, which is the mainstay of the structure of the dioid and the ring (see Figure 1.2). The fundamental difference is the existence of symmetry for the first operation $(a + (-a) = (-a) + a = 0)$ of the rings that is replaced by idempotence $(a + a = a)$ for the first operation (maximum for the (max, +) algebra) of the dioids. This difference is crucial as it is proved that the set is reduced to the neutral element of the first operation ($-\infty$ for the operation max in the (max, +) algebra) if we assume that the structure presents the two properties. This point shows the fundamental difference between automatic control for continuous systems and discrete event systems. The analogies between these two fields are limited to the common core, which is the structure of the semi-ring, even if the authors working in discrete event systems can find some general principles in automatic control allowing the resolution of some specific problems such as "estimation" and "control" in discrete event systems.

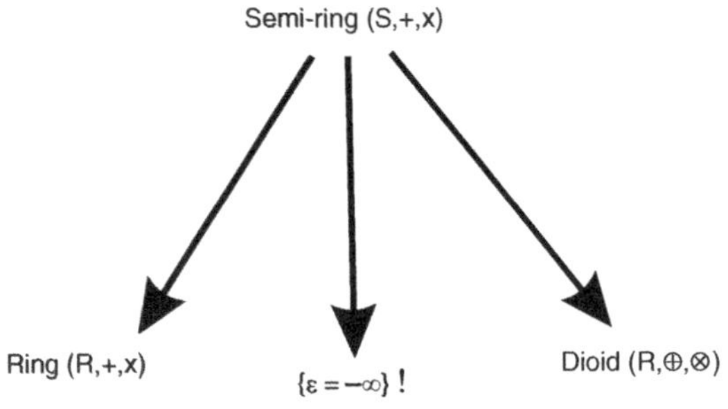

Figure 1.2. *Two algebras: common core and fundamental difference*

1.3.2. *Petri nets*

In the field of discrete event systems, the Petri net allows the modeling of many types of industrial process, transportation system, electronics system, etc. A classical classification of Petri nets can be made with respect to the following two fundamental types (T. Murata gives a good overview of this subject in [MUR 89]).

– The event graphs (each place has exactly one in-going transition and one out-going transition) highlight the synchronization, but no choice is possible. The introduction of time is possible in this model and complex synchronizations can be considered. We will choose this model in the following chapters.

– The state graphs (each transition has exactly one in-going place and one out-going place) consider the choices but do not model the synchronization. The logical aspect is an important characteristic in this model, but the introduction of time is not easy.

– The intersection of the two classes is not empty but only contains Petri nets that do not express either the synchronization or the choice. Therefore, event graphs and state graphs are two fundamental models that equally determine two research fields.

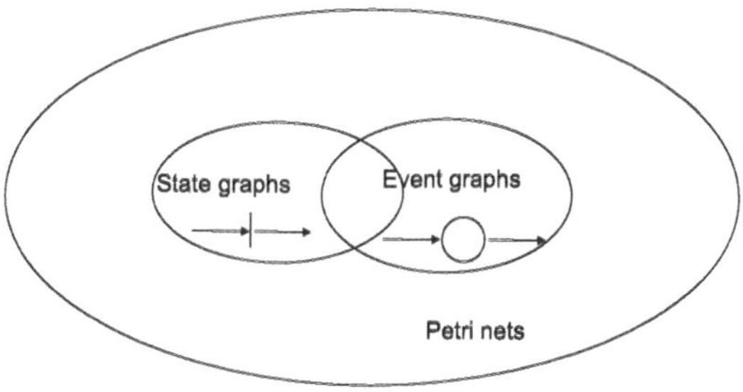

Figure 1.3. *Key structures in Petri nets*

1.3.3. *Time and algebraic models*

All processes naturally present a time evolution and the introduction of time in Petri nets constitutes an important motivation. There are also many everyday examples such as the arrival of students to a course (see Chapter 4 on control). However, the direct introduction of time in general Petri nets quickly leads to a combinatorial blow up of the state-space, which reduces the application field. Contrary to Model Checking, which generates an origin at each event (firing transition) in complex systems, we consider below a unique origin time. Moreover, we will consider specific time models where we can establish a complete algebraic model.

In the field of Petri nets, P-time event graphs, T-time event graphs and time stream event graphs allow us to describe complex synchronizations and to model processes in the production industry, microcircuit design, transportation systems, the food industry or multimedia under the time aspect. The state trajectory is not determined by the iteration of a state equation but is described by a space evolution expressed by the iteration of an interval. The bounds are defined by functions using the minimum, maximum and addition operations. These models are called "interval models". The models obtained are not only (max, +) models but also (min, max, +) models.

Note that the choice of analysis of these models comes from the choice of the use of the "dater", that is, to describe each event by a numbered date or to note by a "counter", i.e. the number of events at each moment. This choice affects the initial modeling, which is a delicate step as it can lead to an unusable model. Note that it is always possible to numerically convert a trajectory expressed with daters into counters, and vice versa. A consequence is that the following chapters will describe *complex* phenomena of *synchronization* for Petri nets presenting *simple structures*. The dual remark is symmetrical: the choice of the counters leads to the analysis of Petri nets presenting *complex structures, weights on the arcs but simple synchronizations.*

From a mathematical point of view, we consider the real space and use the following algebras, that is ($\mathbb{R} \cup \{-\infty\}$, max, $+$), and ($\mathbb{R} \cup \{-\infty\}$, min, max, $+$), possibly completed with $+\infty$. On the other hand, the counters are defined over the integers and not the real numbers. In that case, we consider algebra as $(\mathbb{Z} \cup \{+\infty\}, min, +)$ or $(\mathbb{N} \cup \{+\infty\}, min, +)$ possibly completed with $-\infty$. Remember that the integer linear programming problems need specific techniques.

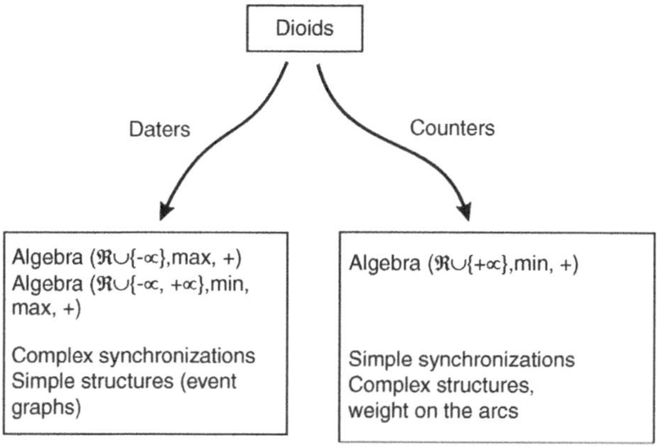

Figure 1.4. *Daters and counters*

1.4. Organization of the book

Figure 1.5 presents the chapters of the book focusing on time models based on daters. We consider the liveness problem as a bad modeling or incorrect parameters which can lead to the absence of a time evolution. The cycle time is the subject of Chapter 3 as it gives an interesting picture of the efficiency of the system. The aim is also to control the process and this point is described in the final chapters.

A comparison of the different classes of time event graphs will be presented at the beginning of chapters on consistency and control as our main interest is on the algebraic models that can describe a broad spectrum of Petri nets. In this book, we also focus on the efficiency of the proposed algorithms: different discussions about complexity and CPU time will be given in the chapters on liveness and predictive control. We also describe some pedagogical examples that will illustrate the main concepts and suggest some new ideas. Each chapter is almost autonomous and the reader only needs to read the preliminary remarks of the chapter on consistency. The book ends with a bibliography and a list of frequently used symbols.

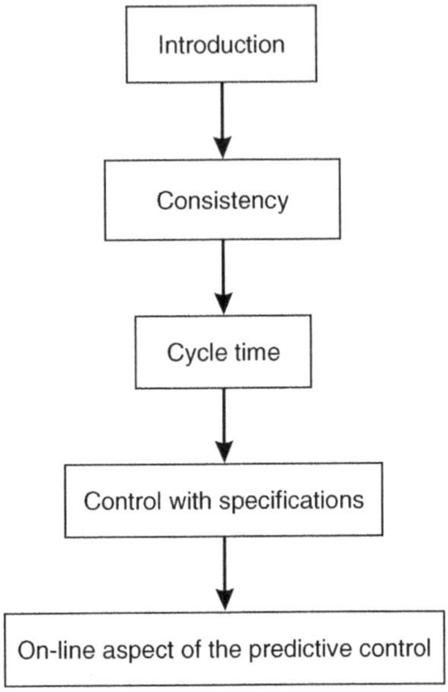

Figure 1.5. *Chapters of the book*

2

Consistency

In this chapter, we consider the (max, +) model of P-time event graphs whose behaviors are defined by lower- and upper-bound constraints. The extremal trajectories of the system starting from an initial interval are characterized by a particular series of matrices for a given finite horizon. Two dual polynomial algorithms are proposed to check the existence of feasible trajectories. The series of matrices are used in the determination of the maximal horizon of consistency and the calculation of the date of the first token deaths.

2.1. Introduction

2.1.1. *Models*

Petri nets (PNs) with time can express the time behavior of discrete event systems with their specifications. Two main behaviors of the transitions can be distinguished: firing as soon as possible in timed PNs and firing in given time intervals for time PNs. Time can be associated with places, transitions and arcs of the PNs. In time stream PNs, temporal intervals are associated with arcs out-going from places and the firing interval of transitions is defined by different semantics [COU 96, DIA 97]. For timed PNs, durations can be associated with places (P-timed PNs) or transitions (T-timed PNs) and the relevant subclasses are equivalent. For time PNs, temporal intervals can similarly be associated with places or transitions but the corresponding subclasses (P-time PNs and T-time PNs) are fundamentally different. In time PNs, a temporal interval of firing is associated with each transition enabled by the marking while a temporal interval of availability is associated with each token that enters a place in P-time PNs (Chapter 6 in [DIA 97]).

Moreover, T-time PNs and P-time PNs correspond to two different semantics of time stream PNs: a T-time PN is similar to a time stream PN for specific intervals, where the semantic of the transitions is “Weak-And”; and the algebraic model of a P-time event graph is identical to a time stream event graph, where the semantic of the transitions is not “Weak-And” but “And” [DID 05].

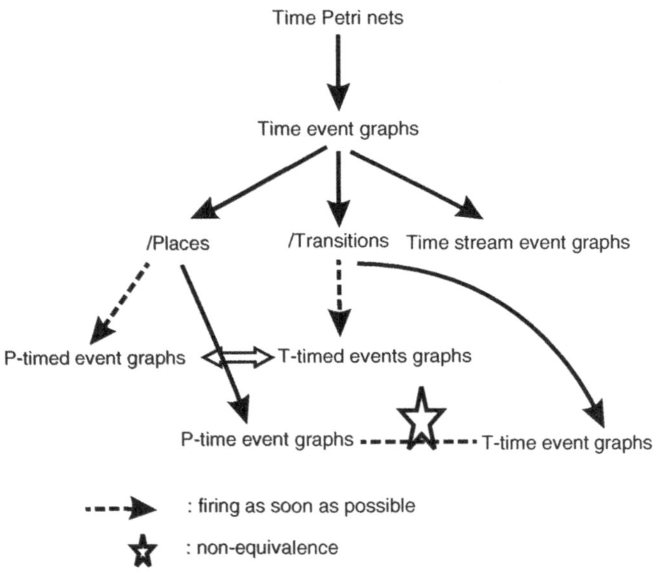

Figure 2.1. *Time Petri nets*

An algebraic summary of the different classes of time event graph is as follows. Time event graphs can be described by a new class of systems called interval systems [DEC 05a] for which the time evolution is not strictly fixed but belongs to intervals. For interval systems, lower and upper interval bounds depend on the operations of maximization, minimization and addition, in the general case. The lower- and upper-bound constraints of T-time event graphs are (max, +) functions while the lower- and upper-bound constraints of P-time event graphs are (max, +) and (min, +) functions, respectively. More details on interval systems can be found in Chapter 4.

2.1.2. *Physical point of view*

P-time event graphs define a set of trajectories that follows the specifications given by the model for a nominal behavior. Firing dates of the transitions belong to the relevant time windows. Indeed, P-time PNs are convenient tools to model manufacturing systems whose operation times must be included between minimum and maximum values. Applications of P-time event graphs can be found in production systems [DEN 05], the food industry [DEC 08] and transportation systems [DEF 08]. A practical example is electroplating lines where the minimum and maximum immersion times guarantee the quality of the chemical treatment [COL 03]: each chemical treatment must be sufficient otherwise the product will not be ready for the next task or treatment; on the other contrary, each chemical treatment must not be too excessive, otherwise the product would be damaged. Other practical examples can be given from the food industry. In good bakery practice, the dough stays in the fermentation room from 3 to 5 hours, the time depending on room temperature and flour or gluten quality. The loaves need to be baked between a minimal and maximal time. If these times are too short or too long (e.g. a synchronization with another operation is not fulfilled), the product will be damaged (bad inner structure and grain in the finished loaf, insufficient or excessive baking). In transportation systems (bus, subway, etc.), we can consider additional constraints such as expected minimum departure frequency, which defines maximum time separation between two vehicle departures at a stop. Moreover, in air traffic, a vehicle must respect a safety distance with the vehicle ahead, which can be expressed by a temporization. In this part, we focus on P-time event graphs [KHA 96] whose evolution can undergo token deaths, which express the loss of resources or parts and failures to meet time specifications.

2.1.3. *Objectives*

However, even if the underlying graph of the event graph is live, the specifications can be too restrictive and some synchronizations cannot follow the desired model, producing losses in resources. A process is composed of machines, resources, etc., and the issue is to know if they can work together following a schedule for a specified period. More particularly, a question is to know whether the different tasks can be sufficiently repeated during a period such that a normal production can be performed without losses of parts. A practical problem might appear when a machine is changed: using a slower/quicker machine can affect the nominal behavior of the complete production line defined by a previous schedule. In this chapter, an objective is to avoid this situation. The relevant notion is the consistency, which can be defined by the existence of a time trajectory following the model. Many approaches, such as control, simulation, optimization, etc., are based on a model and they usually assume that the process follows a normal time evolution expressed by the considered model and not by another model: the correctness of the model is clearly a major problem and it must be checked before the application of any approach.

The natural aim is to characterize the trajectories followed by the system starting from an initial state. In P-time event graphs, it is well known that a simple forward simulation does not guarantee the correct synchronization of the transitions and often leads to token deaths. The first objective is the determination of possible trajectories without token deaths. The concept of extremal (lowest and greatest, see [KUM 95]) trajectories is relevant for the class of P-time event graphs and corresponds to the earliest and latest trajectories. In this part, the objective is to express the *relations* of these extremal trajectories from the model and the initial condition, for a given horizon.

An important notion is the consistency, which can be defined by the existence of a time trajectory following the model. The second question is to know if the different tasks can be sufficiently repeated during a period such that a given production plan can be performed. More precisely, our objective is to know if the different tasks can be repeated infinitely or during a finite period, and to determine the *maximal horizon* (maximal number of events) where the synchronizations of the transitions can be made.

One consequence is that the end of this horizon is also the limit of consistency, which leads to a non-synchronization of the transitions; at least, results in a token death. The last objective is the determination of the *date of the first token deaths*.

In this chapter, an in-depth analysis of P-time event graphs is performed and the algebraic expressions of extremal trajectories are derived. Polynomial algorithms are proposed for the determination of the maximal horizon of temporal consistency and the calculation of the first date of token deaths. Note that the consistency will also be discussed in Chapter 4, in which interval descriptor systems such as time stream event graphs are considered.

Another possible technique is model checking, which is an enumerative method, based on the construction of a state class graph and its analysis. Some authors [BER 91] apply this approach to T-time PNs where each state class is defined by its marking and a set of firing times of the transitions. Starting from a given class, the firing of each enabled transition generates another class, and a procedure establishes the list of different classes and its connections. Generally speaking, model checking faces a combinatorial blow up of the state-space, commonly known as the state explosion problem, even for small systems [MAG 07]: the elementary event graph in the example of Figure 4 in [BER 91], which is composed of two places and two transitions, illustrates this fact. As a consequence, these approaches generally consider a class of models where the graph is finite, that is the time PNs are *bounded* and the bounds of the time intervals are defined in the *rational numbers*. In this chapter, these assumptions are not taken as the considered models are non-bounded event graphs where the values of the temporizations are defined in $\mathbb{R}^+$. Contrary to the polyhedra of the classes, which are generally approximations of the possible dates [BER 07], the spaces considered in this chapter are exact because the concept of lattice is relevant to the event graphs.

In this section, no hypothesis is made on the structure of the event graphs, which need not be strongly connected. The initial marking must only satisfy the classical condition of liveness (no circuit without token), and the usual First In First Out (FIFO) hypothesis for tokens is made.

2.2. Preliminaries

A monoid is a pair $(\mathcal{S}, \oplus)$ where the operation $\oplus$ is associative and presents neutral element ε. A semiring $\mathcal{S}$ is a triple $(\mathcal{S}, \oplus, \otimes)$ where $(\mathcal{S}, \oplus)$ and $(\mathcal{S}, \otimes)$ are monoids, $\oplus$ is commutative, $\otimes$ is distributive in relation to $\oplus$ and the zero element ε of $\oplus$ is the absorbing element of $\otimes(\varepsilon \otimes a = a \otimes \varepsilon = \varepsilon)$. A dioid $\mathcal{D}$ is an idempotent semiring (the operation $\oplus$ is idempotent, that is $a \oplus a = a$). The set $\mathbb{R} \cup \{-\infty\}$, provided with the maximum operation denoted $\oplus$ and the addition denoted $\otimes$ is an example of a dioid denoted $\mathbb{R}_{\max} = (\mathbb{R} \cup \{-\infty\}, \oplus, \otimes)$. The neutral elements of $\oplus$ and $\otimes$ are represented by $\varepsilon = -\infty$ and $e = 0$, respectively. The absorbing element of $\otimes$ is ε. Isomorphic to the previous one by bijection: $x \longmapsto -x$, another dioid is $\mathbb{R} \cup \{+\infty\}$, provided with the minimum operation denoted $\wedge$ and the addition denoted $\odot$. The neutral elements of $\wedge$ and $\odot$ are represented by $T = +\infty$ and $e = 0$, respectively. The absorbing element of $\odot$ is ε. The following conventions are made: $T \otimes \varepsilon = \varepsilon$ and $T \odot \varepsilon = T$. The expressions $a \otimes b$ and $a \odot b$ are identical if at least either a or b is a finite scalar. The partial order denoted $\leqslant$ is defined as follows: $x \leqslant y \Longleftrightarrow x \oplus y = y \Longleftrightarrow x \wedge y = x \Longleftrightarrow x_i \leqslant y_i$, for i from 1 to n in $\mathbb{R}^n$. Notation $x < y$ means that $x \leqslant y$ and $x \neq y$. A dioid $\mathcal{D}$ is complete if it is closed for infinite sums, and the distributivity of the multiplication with respect to addition applies to infinite sums: $(\forall\ c \in \mathcal{D})\ (\forall\ A \subseteq \mathcal{D})\ c \otimes (\bigoplus_{x \in A} x) = \bigoplus_{x \in A} c \otimes x$. For example, $\overline{\mathbb{R}}_{max} = (\mathbb{R} \cup \{-\infty\} \cup \{+\infty\}, \oplus, \otimes)$ is complete. The set of $n.n$ matrices with entries in the complete dioid $\mathcal{D}$ including the two operations $\oplus$ and $\otimes$ is also a complete dioid, which is denoted $\mathcal{D}^{n.n}$. The elements of the matrices in the (max, +) expressions (respectively, (min, +) expressions) are either finite or ε (respectively, T). We can deal with non-square matrices if we complete them with rows or columns provided the entries equal ε (respectively, T). The different operations obey the usual rules of algebra: the notation $\odot$ refers to the multiplication of two matrices in which the $\wedge$-operation is used instead of $\oplus$. The mapping f is said to be residuated if for all $y \in \mathcal{D}$, the least upper bound of subset $\{x \in \mathcal{D} \mid f(x) \leq y\}$ exists and lies in this subset. Mapping $x \in (\overline{\mathbb{R}}_{max})^n \mapsto A \otimes x$, defined over $\overline{\mathbb{R}}_{max}$, is residuated (see [BAC 92]) and the left $\otimes$-residuation of B by A is denoted by: $A\backslash B = \max\{x \in (\overline{\mathbb{R}}_{max})^n \text{ such that } A \otimes x \leqslant B\}$; moreover, $A \setminus B = (-A)^t \odot B$ or $A \odot B = (-A)^t \setminus B$ (see the proof of Theorem 3.21 in part 3.2.3.2 of [BAC 92]) with convention $-\infty - (-\infty) = +\infty$ and $+\infty - (+\infty) = +\infty$.

The Kleene star is defined by: $A^* = \bigoplus_{i=0}^{+\infty} A^i$. Denoted as $G(A)$, an associated graph of square matrix A is deduced from this matrix by associating node i with column i and line i and an arc from node j toward node i with $A_{ij} \neq \varepsilon$. The weight $\mid pa \mid_w$ of path pa is the sum of the labels (weights) on the edges in the path. The length $\mid pa \mid_l$ of path pa is the number of edges in the path. A circuit is a path that starts from and ends at the same node. Using the Kleene star, the two following theorems are dual and will be considered in the dioid of matrices.

THEOREM 2.1 (Theorem 4.75 in part 1 of [BAC 92]).– *Consider equation:*

$$x = A \otimes x \oplus B \qquad [2.1]$$

and inequality:

$$x \geq A \otimes x \oplus B \qquad [2.2]$$

with A and B in complete dioid D. Then, $A^* \otimes B$ is the least solution of [2.1] and [2.2]. □

THEOREM 2.2 (Theorem 4.73 in part 1 of [BAC 92]).– *Consider equation:*

$$x = A\backslash x \wedge B \qquad [2.3]$$

and inequality:

$$x \leq A\backslash x \wedge B \qquad [2.4]$$

with A and B in complete dioid D. Then, $A^*\backslash B$ is the greatest solution of [2.3] and [2.4]. □

For A_{ij} and B_i belonging to $\overline{\mathbb{R}}_{max}$, $A \setminus x \wedge B$ can be written $(-A)^t \odot x \wedge B$. We will need the following result.

COROLLARY 2.1.– Corollary $x \geq A \otimes x \oplus B \Longleftrightarrow x \geq A \otimes x$ and $x \geq A^*B$.

$$x \leq A\backslash x \wedge B \Longleftrightarrow x \leq A\backslash x \otimes x \text{ and } x \leq A^*\backslash B.$$

PROOF.– We only consider the first result. The proof is dual for the second part. First, $\Longrightarrow$: this is an application of Theorem 2.1. Conversely, $\Longleftarrow$: $x \geq A^*B$ implies $x \geq B$. □

We consider below the "dater" description in the (max, +) algebra: each variable $x_i(k)$ represents the date of the kth firing of transition x_i. If we assume a FIFO functioning of the places, which guarantees that the tokens do not overtake one another, a correct numbering of the events can be carried out.

2.3. Models and principle of the approach

In a first section, the definition of P-time event graphs is given and the firing interval of each transition is described. Using the dater form, the algebraic model is built. This model will be analyzed in the next sections. An elementary production system is described and the principle of the approach is presented.

2.3.1. *P-time event graphs*

A PN is a pair (G, M_0), where $G = (S_v, C_o)$ is a bipartite graph with a finite number of nodes (the set S_v), which are partitioned into the disjoint sets of places P and transitions TR (transitions are denoted x while temporization are denoted T, T^- or T^+); C_o consists of pairs of the form (p_i,x_i) and (x_i,p_i) with $p_i \in P$ and $x_i \in TR$. The initial marking M_0 is a vector of dimension $\mid P \mid$ whose elements denote the number of initial tokens in the respective places. The set ${}^{\bullet}p$ is the set of input transitions of p and $p^{\bullet}$ is the set of output transitions of place $p \in P$. The set ${}^{\bullet}x_i$ (respectively, $x_i^{\bullet}$) is the set of input (respectively, output) places of the transition $x_i \in TR$.

For a PN with $|P|$ places and $|TR|$ transitions, the incidence matrix $W = [W_{ij}]$ is a $|P| \times |TR|$ matrix of integers and its entry is given by $W_{ij} = W_{ij}^{+} - W_{ij}^{-}$ where W_{ij}^{+} is the weight of the arc from the transition j to the place i and W_{ij}^{-} is the weight of the arc from the place i to the transition j.

In a PN, a firing sequence from a marking M, implies a string of successive markings. The characteristic vector s of a firing sequence is such that each component is an integer corresponding to the number of firings of the corresponding transition. A marking M reached from initial marking M_0 by firing of a sequence that can be calculated by the fundamental relation: $M = M_0 + W \times s$.

DEFINITION 2.1.– *A PN is called an event graph if each place has exactly one upstream and one downstream transition.*

P-time PNs allow the modeling of discrete event systems with sojourn time constraints of the tokens inside the places. Consistent with the dioid $\overline{\mathbb{R}}_{max}$, we associate a temporal interval defined in $\mathbb{R}^{+} \times (\mathbb{R} \cup \{+\infty\})$ with each place. Each place $p_l \in P$ is associated with an interval $[T_l^-, T_l^+]$, where T_l^- is the lower bound and T_l^+ the upper bound. Its initial marking is denoted $(M_0)_l$ or m_l.

DEFINITION 2.2.– *A P-time event graph is a triplet* (EG, M_0, g) *where EG is an event graph,* M_0 *is the initial marking and the mapping* g *is defined by* $p_l \to [T_l^-, T_l^+]$ *with* $0 \leq T_l^- \leq T_l^+$ *from* P *to* $\mathbb{R}^{+} \times (\mathbb{R}^{+} \cup \{+\infty\})$.

The interval $[T_l^-, T_l^+]$ is the static interval of duration time of a token in place p_l. The token must stay in this place during the minimum residence duration T_l^-. Before this duration, the token is in a state of unavailability to fire the out-going transition. The value T_l^+ is a maximum residence duration after which the token must leave the place p_l (and can contribute to the enabling of the downstream transition). If not, the system falls into a token-dead state. So, the token is available to fire the out-going transition in the time interval $[T_l^-, T_l^+]$.

A consequence is a possible bad synchronization of each transition, which is the out-going transition of more than one place. This situation occurs when the firing dates of the in-going transitions of at least two places are incoherent. This non-synchronization can be solved by a prediction of the evolution of the tokens in the places.

EXAMPLE 2.1.– Let us consider the P-time event graph of Figure 2.2. The initial marking is $(1\ \ 1\ \ 1\ \ 1)^t$ and the temporal intervals are: $[T_1^-, +\infty] = [3, +\infty]$, $[T_2^-, +\infty] = [6, +\infty]$, $[T_3^-, T_3^+] = [1, 2]$ and $[T_4^-, T_4^+] = [3, 11]$. Let us consider the following simulation for $k = 0$, 1 and 2.

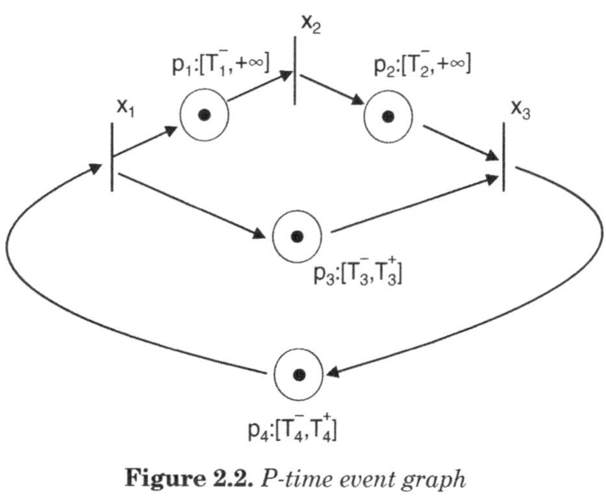

Figure 2.2. *P-time event graph*

The initial marking is $(1 \quad 1 \quad 1 \quad 1)^t$ and the temporal intervals are: $[T_1^-, +\infty] = [3, +\infty]$, $[T_2^-, +\infty] = [6, +\infty]$, $[T_3^-, T_3^+] = [1, 2]$ and $[T_4^-, T_4^+] = [3, 11]$.

k	0	1	2
x_1	4	11	11
x_2	0	7	14
x_3	0	6	13

k	0	1	2
p_1	$[7, +\infty]$	$[14, +\infty]$	$[14, +\infty]$
p_2	$[6, +\infty]$	$[13, +\infty]$	$[20, +\infty]$
p_3	$[5, 6]$	$[12, 13]$	$[12, 13]$
p_4	$[3, 11]$	$[9, 17]$	$[16, 24]$

The first table contains the firing dates while each column k of the second table is the bounds of the sojourn time (in absolute time) of the tokens, produced by the kth firing of the transitions x_1, x_2 and x_3 in each place. We assume that the tokens of the initial marking are available immediately at $k = 0$. Let us consider the firing of transition x_3 for $k = 3$, which needs to use the tokens present in its upstream places p_2 and p_3 produced at $k = 2$. However, this synchronization does not occur because the interval $[20, +\infty] \cap [12, 13]$ is empty. A consequence is the death of the token in place p_3 at time $t = 13$.

However, transition x_1 can be fired at $t = 18$ because the interval of sojourn time of the token in place p_4 is $[16, 24]$. Therefore, a token is added in place p_3 with time interval $[19, 20]$ and the firing of transition x_3 can occur at time $t = 20$ because the interval $[20, +\infty] \cap [12, 13]$ is replaced by the interval $[20, +\infty] \cap [19, 20]$.

Now, we consider the dater form, which will give an efficient description.

2.3.2. *Dater form*

In an event graph, card$({}^\bullet p_l)$ = card$(p_l^\bullet) = 1$ for each place $p_l \in P$ and we can associate only a pair (x_i, x_j) with each place $p_l \in P$, such that transition x_j is in-going ($x_j \in^\bullet p_l$) and transition x_i is out-going ($x_i \in p_l^\bullet$). Time interval $[T_l^-, T_l^+]$ and initial marking m_l are also associated with place p_l. The evolution of the P-time event graph is described by the following inequalities expressing relations between the firing dates of transitions:

$\forall p_l \in P$ with $x_j \in^\bullet p_l$ and $x_i \in p_l^\bullet, T_l^- + x_j(k - m_l) \leq x_i(k)$ and $x_i(k) \leq T_l^+ + x_j(k - m_l)$

Now, let us consider a pair of transitions (x_i, x_j) and a given marking q. These conditions define a set of places $P_{i,j,q}$, which can be empty or contain more than one place: each place p_l of $P_{i,j,q}$ satisfies $\{x_j\} =^\bullet p_l$, $\{x_i\} = p_l^\bullet$ and $m_l = q$. $\forall p_l \in P_{i,j,q}$, we can take the maximum of lower bounds T_l^- and the minimum of upper bounds T_l^+ and we denote the corresponding values $T_{i,j,q}^- \in \mathbb{R}^+$ and $T_{i,j,q}^+ \in \mathbb{R}^-$. More formally, $T_{i,j,q}^- = \bigoplus_{\forall p_l \in P_{i,j,q}} T_l^-$ and $T_{i,j,q}^+ = \bigwedge_{\forall p_l \in P_{i,j,q}} T_l^+$.

REMARK 2.1.– Naturally, if for each pair of transitions (x_i, x_j), there is a unique place $p_l \in P$ in the event graph, we can simplify the notation and replace $a_{i,j,q}^-$ by $a_{i,j}^-$ and $a_{i,j,q}^+$ by $a_{i,j}^+$. In the figures of the chapter, each temporization is directly indexed with the index l of the relevant place p_l.

Therefore, the system can be described as follows:

$\forall P_{i,j,q} \subset P, T_{i,j,q}^- + x_j(k-q) \leq x_i(k)$ and $x_i(k) \leq T_{i,j,q}^+ + x_j(k-q)$

After permutation of indexes i and j, and application of simple transformations, the latter inequality is equivalent to $-T_{j,i,q}^+ + x_j(k+q) \leq x_i(k)$.

In short, $T^-_{i,j,q}+x_j(k-q) \leq x_i(k)$ and $-T^+_{j,i,q}+x_j(k+q) \leq x_i(k)$.

The system can now be expressed with matrices in (max, +) algebra. This allows the writing of a synthetic description on a horizon defined by the maximal initial marking $m = \bigoplus_{\forall p_k \in P} m_{p_k} = \bigoplus_{\forall P_{i,j,q} \subset P} q$.

$$x(k) \geq \bigoplus_{0 \leq q \leq m} A^-_q \otimes x(k-q) \qquad [2.5]$$

$$x(k) \geq \bigoplus_{0 \leq q \leq m} A^+_q \otimes x(k+q) \qquad [2.6]$$

with $(A^-_q)_{ij} = T^-_{i,j,q}$ if $T^-_{i,j,q}$ exists in $\mathbb{R}$ or ε otherwise,

$(A^+_q)_{ij} = -T^+_{j,i,q}$ if $T^+_{j,i,q}$ exists in $\mathbb{R}$ or ε otherwise.

For instance, in Figure 2.3, $(A^-_0)_{3,1} = T^-_{3,1,0} = T^-_{p3} = T^-_3$ and $(A^+_0)_{1,3} = -T^+_{3,1,0} = -T^+_{p3} = -T^+_3$ for place p_3.

This model completely describes the relevant P-time event graph by giving a lower bound of state $x(k)$. This lower bound depends on values $x(k-q)$ and $x(k+q)$ for $q = 0$ to m in inequalities [2.5] and [2.6], respectively. As inequality [2.5] corresponds to a classical timed event graph (*without assumption of earliest functioning*), a P-time event graph can be seen as a timed event graph [2.5] following additional specifications [2.6]. The timed event graph can express the physical limitations of the process as the minimal cooking time while the upper bounds describe quality criteria on the finished parts and products: the respect of these constraints needs an anticipation of the future behavior of the process. Therefore, the calculation of the lower-bound trajectory cannot be made from only the past trajectory like a timed event graph working in the earliest functioning, but must use a prediction of the future evolution. In the following parts, we will see that this remark also holds for the upper bound.

REMARK 2.2.– Some authors add the additional assumption of the earliest behavior and replace the inequality in [2.5] with an equality. Therefore, they limit the possibility of modeling of P-time event graphs, which does not describe a unique trajectory but a set of trajectories. Particularly, P-time event graphs can describe uncertainties on the sequence time of the process [COL 03, DEN 05], while the minimal and maximal times of each task are exactly known. For instance, the choice of a cooking time in the middle of interval $[T_{\min}, T_{\max}]$ guarantees the quality of the final product but other choices are possible. The choices $T_{\min}$ and $T_{\max}$ are risky (underdone, overdone) as the parameters of the oven can change.

The aim of this chapter was the analysis of the implicit model defined by [2.5] and [2.6]. Before considering the general model, we first introduce the principle of the general approach with a simple example. This section only uses commonly used algebra. A more general study will be given in the following section.

2.3.3. *Principle of the approach (example 2)*

Let us consider an elementary production system composed of two parallel lines that start at the same time. The process is described by a P-time event graph in Figure 2.3. The first line is composed of two tasks while the second only corresponds to the cooking of a product (T_3^-). The tasks of line 1 are successively the making of a packet (T_1^-) and its moving (T_2^-). When the activities are completed, the finished product is packed (T_4^-). Naturally, the cooking time T_3^+ must not be too excessive, otherwise the product will be damaged ($x_3 > T_3^+ + x_1$).

The following inequalities describe the two lines. $T_1^- + x_1 \leq x_2$, $T_2^- + x_2 \leq x_3$ and $T_3^- + x_1 \leq x_3 \leq T_3^+ + x_1$.

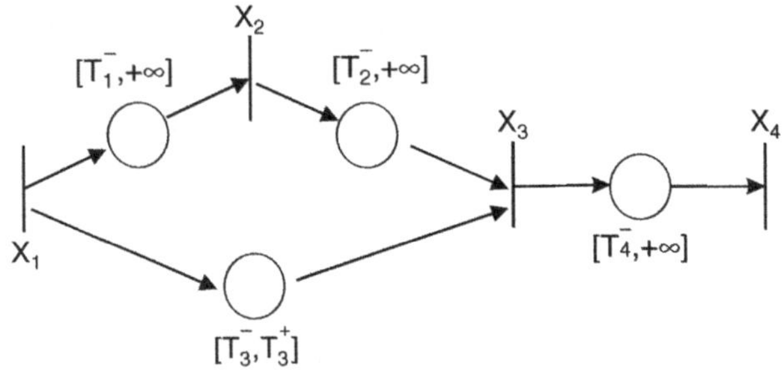

Figure 2.3. *Principle*

Therefore, $T_1^- + T_2^- + x_1 \leq x_3$ for line 1, and $T_3^- + x_1 \leq x_3 \leq T_3^+ + x_1$ for line 2. So, $T_1^- + T_2^- + x_1 \leq x_3 \leq T_3^+ + x_1$ and, consequently, condition $T_1^- + T_2^- \leq T_3^+$ is necessary, otherwise the system is inconsistent. In other words, the process does not work if time T_3^+ is less than the sum of the temporizations T_1^- and T_2^-.

Another explanation is as follows. Inequality $x_3 \leq T_3^+ + x_1$ can be written $-T_3^+ + x_3 \leq x_1$. From $T_1^- + x_1 \leq x_2$; $T_2^- + x_2 \leq x_3$; $-T_3^+ + x_3 \leq x_1$, we can deduce that $-T_3^+ + T_2^- + T_1^- + x_1 \leq x_1$. A necessary and sufficient condition of existence of a solution is $-T_3^+ + T_2^- + T_1^- \leq 0$.

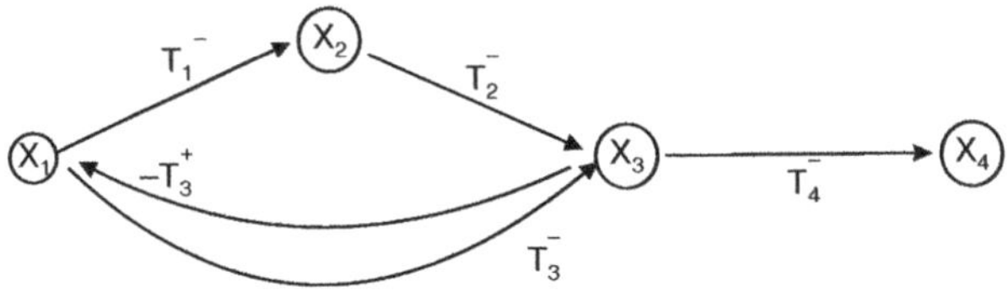

Figure 2.4. *Associated graph 1*

These inequalities can be described by a graph (see Figure 2.4) defined as follows. The vertices correspond to the transitions of the PNs and a directed arc from j to i is associated with each inequality $a + x_j \leq x_i$ where $a = T^-$ or $a = -T^+$.

This graph shows that the term $-T_3^+ + T_2^- + T_1^-$ is the weight of the circuit defined by transitions x_1, x_2, x_3 and x_1. We can say that the system will be consistent if any circuit of the graph has a negative or null weight. In this case, a sequence of firing dates can be found. If the process follows these dates, the production will be satisfactory as each cooked product never stays in oven after delay T_3^+.

Now, we generalize this first intuitive study.

2.4. Analysis in the "static" case

Let us assume that the process is static or, in other words, that the marking is null: $m = 0$. Therefore, the model described by [2.5] and [2.6] is reduced to the following form.

$$x \geq (A_0^- \oplus A_0^+) \otimes x \qquad [2.7]$$

Inequalities of this form are classical in the (max, +) context. The following well-known result clearly shows that the consistency analysis of [2.7] needs an analysis of the circuits in the static case.

PROPOSITION 2.1.– There is a finite vector $x \in \mathbb{R}^{|x|}$ satisfying [2.7] if and only if the associated graph of matrix $A_0^- \oplus A_0^+$ has only circuits with non-positive weight.

Recall that $(A^*)_{i,i}$ is the greatest weight of the circuits going by vertex i of the associated graph of matrix A. Another formulation of the proposition is that a necessary and sufficient condition for the existence of a state in $\mathbb{R}$ (not in $\mathbb{R}_{\max}$) is that $((A_0^- \oplus A_0^+)^*)_{i,i}$ converges on $\mathbb{R}_{\max}$ and not on $T = +\infty$ for any index i.

The following form makes the connection with the study in [VAN 05]. First, let us note that $x \geq A_0^+ \otimes x$ is equivalent to $x \leq A_0^+ \backslash x$. So, inequality $x \geq (A_0^- \oplus A_0^+) \otimes x$ is equivalent to $A_0^- \otimes x \leq x \leq A_0^+ \backslash x$. This system implies the following expression

$$A_0^- \otimes x \leq A_0^+ \backslash x = (-A_0^+)^t \odot x \qquad [2.8]$$

which has been analyzed [VAN 05] in the proposition below. It is given with a slightly modified notation.

PROPOSITION 2.2.– [VAN 05] There is a finite vector $x \in \mathbb{R}^{|x|}$ satisfying $A_0^- \otimes x \leq (-A_0^+)^t \odot x$ if and only if the associated graph of $A_0^+ \otimes A_0^-$ contains circuits with only non-positive weight.

Note that the relation defined by [2.8] has been deduced from [2.7] or, in other words, the set defined by [2.8] includes the set defined by [2.7]. However, the relations are not mathematically equivalent as shown in the following counterexample.

EXAMPLE 2.2.– $A_0^- = \begin{pmatrix} -10 & -20 \\ -15 & 0 \end{pmatrix}$ and $(-A_0^+)^t = \begin{pmatrix} 1 & 1 \\ 1 & 1 \end{pmatrix}$ $x = \begin{pmatrix} 11 \\ 1 \end{pmatrix}$

$A_0^- \otimes x = \begin{pmatrix} 1 \\ 1 \end{pmatrix} \leq (-A_0^+)^t \odot x = \begin{pmatrix} 2 \\ 2 \end{pmatrix}$ but $\begin{pmatrix} 1 \\ 1 \end{pmatrix} \leq \begin{pmatrix} 11 \\ 1 \end{pmatrix} \nleq \begin{pmatrix} 2 \\ 2 \end{pmatrix}$

The following results allow a comparison of the consistency of [2.7] and [2.8], based on the circuits of the associated graphs. The usual multiplication is denoted by a mid-dot below.

PROPOSITION 2.3.– $(\forall \mathbf{k} \in \mathbb{N})(A \oplus B)^{2 \cdot k} \geq (A \otimes B)^k$

PROOF.– The inductive proof is as follows. The hypothesis is H_k: $(A \oplus B)^{2 \cdot k} \geq (A \otimes B)^k$.

Initial step. H_1 defined by $(A \oplus B)^2 \geq (A \otimes B)^1$ is true as $(A \oplus B)^2 = A^2 \oplus A \otimes B \oplus B \otimes A \oplus B^2$

Inductive step. Let us assume that H_k is true for a given k$\in \mathbb{N}$. We now prove that the formula is true for $k+1$.

$(A \oplus B)^{2.(k+1)} = (A \oplus B)^{2 \cdot k} \otimes (A \oplus B)^2 \geq (A \otimes B)^k \otimes (A \otimes B)^1 = (A \otimes B)^{k+1}$ and H_{k+1} is proved. □

PROPOSITION 2.4.– $(A \oplus B)^* \geq (A \otimes B)^*$

PROOF.– By definition, $(A \oplus B)^* = \bigoplus_{i \in \mathbb{N}} (A \oplus B)^i \geq \bigoplus_{k \in \mathbb{N}} (A \oplus B)^{2 \cdot k}$.

The previous proposition implies that $\bigoplus_{k \in \mathbb{N}} (A \oplus B)^{2 \cdot k} \geq \bigoplus_{k \in \mathbb{N}} (A \otimes B)^k = (A \otimes B)^*$ and the proposition is proved. □

Therefore, the application of the previous result gives: $(A_0^- \oplus A_0^+)^* = (A_0^- \oplus A_0^+)^* \oplus (A_0^+ \oplus A_0^-)^* \geq (A_0^- \otimes A_0^+)^* \oplus (A_0^+ \otimes A_0^-)^*$. Particularly, $((A_0^- \oplus A_0^+)^*)_{i,i} \geq ((A_0^- \otimes A_0^+)^*)_{i,i} \oplus ((A_0^+ \otimes A_0^-)^*)_{i,i}$.

Consequently, even if $\forall i \in ((A_0^+ \otimes A_0^-)^*)_{i,i} \leq 0$, the term $((A_0^- \oplus A_0^+)^*)_{i,i}$ can be positive. Therefore, if the associated graph of $(A_0^+ \otimes A_0^-)$ contains circuits with only non-positive weight (Proposition 2.2 [VAN 05]), the associated graph of matrix $(A_0^- \oplus A_0^+)$ can have circuits with positive weights. Therefore, application of the previous two propositions shows that inequality [2.7] can be inconsistent while inequality [2.8] is consistent.

In conclusion, this part shows that the consistency depends on the circuits in an associated graph. This analysis will now be generalized to an arbitrary initial marking. The implicit model described by [2.5] and [2.6] will first be rewritten on a short horizon in order to simplify the analysis. Then, this new form will be used to calculate extremal trajectories and to analyze the consistency in the following sections.

2.5. "Dynamic" model

Now, we consider an arbitrary initial marking. The following proposition is about the existence of a state trajectory in $\mathbb{R}$ (and not in $\mathbb{R}_{\max}$).

PROPOSITION 2.5.– A necessary condition for the existence of a state trajectory in $\mathbb{R}$ is that the associated graph of matrix $A_0^- \oplus A_0^+$ has only circuits with non-positive weight.

PROOF.– From inequalities [2.5] and [2.6] of the model, we have:

$$\begin{aligned} x(k) &\geq \bigoplus_{0\leq q\leq m} A_q^- \otimes x(k-q) \oplus \bigoplus_{0\leq q\leq m} A_q^+ \otimes x(k+q) \\ &= (A_0^- \oplus A_0^+) \otimes x(k) \oplus \bigoplus_{1\leq q\leq m} A_q^- \otimes x(k-q) \\ &\quad \oplus \bigoplus_{1\leq q\leq m} A_q^+ \otimes x(k+q) \end{aligned} \qquad [2.9]$$

We can deduce that $x(k) \geq (A_0^- \oplus A_0^+) \otimes x(k)$ and apply Proposition 2.1. □

From [2.9], we deduce the following inequalities, where the right-hand term of the first inequality represents the least solution of [2.9].

$$\begin{cases} x(k) \geq (A_0^- \oplus A_0^+)^* \\ \quad \otimes[\bigoplus_{1\leq q\leq m} A_q^- \otimes x(k-q) \oplus \bigoplus_{1\leq q\leq m} A_q^+ \otimes x(k+q)] \\ x(k) \geq (A_0^- \oplus A_0^+) \otimes x(k) \end{cases} \qquad [2.10]$$

The following property shows that [2.10] completely expresses the model.

PROPOSITION 2.6.– The inequalities [2.10] and the implicit model defined by [2.5] and [2.6] are equivalent.

PROOF.– Immediate from Corollary 2.1. □

Now, let us introduce the following notations.

$$\begin{cases} \mathbb{A}_0^{=} = A_0^- \oplus A_0^+ \\ \mathbb{A}_q^- = (A_0^- \oplus A_0^+)^* \otimes A_q^-, \text{ for } q = 1 \text{ to } m. \\ \mathbb{A}_q^+ = (A_0^- \oplus A_0^+)^* \otimes A_q^+, \text{ for } q = 1 \text{ to } m. \end{cases}$$

Therefore, model [2.10] can be rewritten as follows.

$$\begin{cases} x(k) \geq \mathbb{A}_0^{=} \otimes x(k) \\ x \geq \bigoplus_{1 \leq q \leq m} \mathbb{A}_q^- \otimes x(k-q) \oplus \bigoplus_{1 \leq q \leq m} \mathbb{A}_q^+ \otimes x(k+q) \end{cases} \quad [2.11]$$

System [2.11] can be simplified by defining an augmented state vector. We propose below one possible technique. The new state vector, denoted $\mathcal{X}$, includes variables $x(k)$, $x_i^-(k)$ and $x_i^+(k)$, for $i = 1$ to $m-1$ defined as follows.

$x_1^-(k) = x(k-1), x_1^+(k) = x(k+1), x_i^-(k) = x_{i-1}^-(k-1)$ and $x_i^+(k) = x_{i-1}^+(k+1)$, for $i = 2$ to $m-1$.

$\mathcal{X} = ((x_{m-1}^-)^t \ \dots \ (x_2^-)^t \ (x_1^-)^t \ (x)^t \ (x_1^+)^t \ (x_2^+)^t \ \dots \ (x_{m-1}^+)^t)^t$ (t: transpose). The dimension of $\mathcal{X}$ is denoted n, which is equal to the product of the dimension of x by $2.m-1$. The following inequalities completely describe both static part and dynamic part of the system.

$x_1^-(k) = x(k-1) \Leftrightarrow x_1^-(k) \geq x(k-1)$ and $x(k) \geq x_1^-(k+1)$

$x_1^+(k) = x(k+1) \Leftrightarrow x_1^+(k) \geq x(k+1)$ and $x(k) \geq x_1^+(k-1)$

$x_i^-(k) = x_{i-1}^-(k-1) \Leftrightarrow x_i^-(k) \geq x_{i-1}^-(k-1)$ and $x_{i-1}^-(k) \geq x_i^-(k+1)$, for $i = 2$ to $m-1$.

$x_i^+(k) = x_{i-1}^+(k+1) \Leftrightarrow x_i^+(k) \geq x_{i-1}^+(k+1)$ and $x_{i-1}^+(k) \geq x_i^+(k-1)$, for $i = 2$ to $m-1$.

Finally, the simplified inequalities are as follows:

$$\begin{cases} \mathcal{X}(k) \geq \mathcal{A}^{=} \otimes \mathcal{X}(k) \\ \mathcal{X}(k) \geq \mathcal{A}^{-} \otimes \mathcal{X}(k-1) \\ \mathcal{X}(k) \geq \mathcal{A}^{+} \otimes \mathcal{X}(k+1) \end{cases} \qquad [2.12]$$

or

$$\begin{pmatrix} \mathcal{X}(k) \\ \mathcal{X}(k+1) \end{pmatrix} \geq \begin{pmatrix} \mathcal{A}^{=} & \mathcal{A}^{+} \\ \mathcal{A}^{-} & \mathcal{A}^{=} \end{pmatrix} \otimes \begin{pmatrix} \mathcal{X}(k) \\ \mathcal{X}(k+1) \end{pmatrix} \qquad [2.13]$$

We will now illustrate the procedure for the synthesis of matrices $\mathcal{A}^{=}$, $\mathcal{A}^{-}$ and $\mathcal{A}^{+}$ in inequalities [2.12]. Let $m = 3$. So, $\mathcal{X} = \begin{pmatrix} (x_2^-)^t & (x_1^-)^t & (x)^t & (x_1^+)^t & (x_2^+)^t \end{pmatrix}^t$

$x(k) \geq \mathbb{A}_3^- \otimes x_2^-(k-1) \oplus \mathbb{A}_2^- \otimes x_1^-(k-1) \oplus \mathbb{A}_1^- \otimes x(k-1) \oplus (\mathbb{A}_0^- \oplus \mathbb{A}_0^+) \otimes x(k) \oplus \mathbb{A}_1^+ \otimes x(k+1) \oplus \mathbb{A}_2^+ \otimes x_1^+(k+1) \oplus \mathbb{A}_3^+ \otimes x_2^+(k+1)$

$x_1^-(k) = x(k-1) \Leftrightarrow x_1^-(k) \geq x(k-1)$ and $x(k) \geq x_1^-(k+1)$

$x_1^+(k) = x(k+1) \Leftrightarrow x_1^+(k) \geq x(k+1)$ and $x(k) \geq x_1^+(k-1)$

$x_2^-(k) = x_1^-(k-1) \Leftrightarrow x_2^-(k) \geq x_1^-(k-1)$ and $x_1^-(k) \geq x_2^-(k+1)$

$x_2^+(k) = x_1^+(k+1) \Leftrightarrow x_2^+(k) \geq x_1^+(k+1)$ and $x_1^+(k) \geq x_2^+(k-1)$

$$\mathcal{A}^{=} = \begin{pmatrix} \varepsilon & \varepsilon & \varepsilon & \varepsilon & \varepsilon \\ \varepsilon & \varepsilon & \varepsilon & \varepsilon & \varepsilon \\ \varepsilon & \varepsilon & \mathbb{A}_0^{=} & \varepsilon & \varepsilon \\ \varepsilon & \varepsilon & \varepsilon & \varepsilon & \varepsilon \\ \varepsilon & \varepsilon & \varepsilon & \varepsilon & \varepsilon \end{pmatrix}, \qquad \mathcal{A}^{-} = \begin{pmatrix} \varepsilon & I & \varepsilon & \varepsilon & \varepsilon \\ \varepsilon & \varepsilon & I & \varepsilon & \varepsilon \\ \mathbb{A}_3^- & \mathbb{A}_2^- & \mathbb{A}_1^- & I & \varepsilon \\ \varepsilon & \varepsilon & \varepsilon & \varepsilon & I \\ \varepsilon & \varepsilon & \varepsilon & \varepsilon & \varepsilon \end{pmatrix},$$

$$\text{and } \mathcal{A}^{+} = \begin{pmatrix} \varepsilon & \varepsilon & \varepsilon & \varepsilon & \varepsilon \\ I & \varepsilon & \varepsilon & \varepsilon & \varepsilon \\ \varepsilon & I & \mathbb{A}_1^+ & \mathbb{A}_2^+ & \mathbb{A}_3^+ \\ \varepsilon & \varepsilon & I & \varepsilon & \varepsilon \\ \varepsilon & \varepsilon & \varepsilon & I & \varepsilon \end{pmatrix}$$

Also, the diagonal of $\mathcal{A}^-$ can be modified such that the non-decrease of the state trajectory is guaranteed. This operation keeps expressions $x_1^-(k) = x(k-1)$, $x_2^-(k) = x_1^-(k-1)$, ... unchanged, because $x_2^-(k) \geq x_1^-(k-1) \oplus x_2^-(k-1) = x_1^-(k-1) \oplus x_1^-(k-2) = x_1^-(k-1)$, for instance.

These expressions describe the "lower" constraints on $\mathcal{X}$ produced by the model, which can maximize the state estimation. The set of inequalities [2.12] clearly describes a forward part ($\mathcal{X}(k) \geq \mathcal{A}^- \otimes \mathcal{X}(k-1)$, a backward part and a "static" (i.e neither backward, nor forward) part ($\mathcal{X}(k) \geq \mathcal{A}^= \otimes \mathcal{X}(k)$). These relations lead to complex backward/forward interconnections that can produce inconsistencies in the model.

Symmetrically, as mapping $\mathcal{A}^= \otimes \mathcal{X}(k)$, $\mathcal{A}^- \otimes \mathcal{X}(k-1)$ and $\mathcal{A}^+ \otimes \mathcal{X}(k+1)$ are residuated, the application of property denoted f3 (in [BAC 92], part 4.4.4) gives the following form: it expresses every "upper" constraint on $\mathcal{X}(k)$, which can minimize it.

$$\begin{cases} \mathcal{X}(k) \leq \mathcal{A}^= \backslash \mathcal{X}(k) \\ \mathcal{X}(k) \leq \mathcal{A}^- \backslash \mathcal{X}(k+1) \\ \mathcal{X}(k) \leq \mathcal{A}^+ \backslash \mathcal{X}(k-1) \end{cases} \qquad [2.14]$$

Each model can be deduced from the other one by duality and each lower (upper) matrix, respectively, corresponds to an upper (lower) matrix with the same notation: symbols $\geq$, $\oplus$ and $\otimes$, respectively, correspond to $\leq$, $\wedge$ and $\backslash$; number of events $k-1$ is replaced by $k+1$ and conversely.

In the following section, the time evolution of model [2.12] is analyzed.

2.6. Extremal acceptable trajectories by series of matrices

Unlike the class of timed event graphs, which define a unique trajectory on assumption of earliest behavior, P-time event graphs define a set of trajectories that depend on matrices $\mathcal{A}^{=}$, $\mathcal{A}^{-}$ and $\mathcal{A}^{+}$. The aim of this section is the determination of the lowest (respectively, greatest) acceptable trajectories satisfying an initial condition given by $\mathcal{X}(0) \in [\mathcal{X}_0^-, \mathcal{X}_0^+]$. In the following text, it is shown that the existence of a trajectory depends on special new matrices denoted w_k.

As a finite horizon, $h \in \mathbb{N}$ is considered, a realistic assumption is that the model operates on the same horizon: we consider the model [2.13] for $k = 0, ..., h-1$.

2.6.1. *Lowest state trajectory*

The following algorithms give the lowest and greatest trajectories satisfying the objective. The first step (a) is the forward calculation of parameters w_k, which only depend on the model. Starting from the initial condition $\mathcal{X}_0^-$ (respectively, $\mathcal{X}_0^+$), the second step (b) is also based on a forward iteration. It expresses a first estimate of the lowest (respectively, greatest) trajectory denoted β_k^- (respectively, β_k^+), which is finally improved by a maximization (respectively, a minimization) in step (c). The final result is the lowest (respectively, greatest) trajectory denoted by $\mathcal{X}_k^-$ (respectively, $\mathcal{X}_k^+$).

THEOREM 2.3.– *If the process operates on horizon $h \in \mathbb{N}$ and if matrices w_k defined below have no positive circuit, the lowest state trajectory in $\mathbb{R} \cup \{-\infty\}$ satisfying $\mathcal{X}(0) \geq \mathcal{X}_0^- \in (\mathbb{R} \cup \{-\infty\})^n$ is given by the following forward / backward algorithm.*

Forward/backward algorithm

a) Coefficients of w_k by forward iteration

Initialization: $w_0 = \mathcal{A}^{=}$

for $k = 1$ to h, $\quad w_k = \mathcal{A}^{=} \oplus \mathcal{A}^{-} \otimes (w_{k-1})^* \otimes \mathcal{A}^{+}$

b) First estimate β_k^- by forward iteration

Initialization: $\beta_0^- = \mathcal{X}_0^-$

for $k = 1$ *to* h, $\beta_k^- = \mathcal{A}^- \otimes (w_{k-1})^* \otimes \beta_{k-1}^-$,

c) Trajectory $\mathcal{X}_k^-$ by backward iteration

Initialization: $\mathcal{X}_h^- = (w_h)^* \otimes \beta_h^-$

for $k = h-1$ *to* 0, $\mathcal{X}_k^- = (w_k)^* \otimes [\mathcal{A}^+ \otimes \mathcal{X}_{k+1}^- \oplus \beta_k^-]$

PROOF.– Theorem 2.1 shows that the smallest solution satisfying $\mathcal{X}(k) \geq \mathcal{A}^= \otimes \mathcal{X}(k) \oplus \mathcal{A}^- \otimes \mathcal{X}(k-1) \oplus \mathcal{A}^+ \otimes \mathcal{X}(k+1)$ with $\mathcal{X}(0) \geq \mathcal{X}_0^-$ also satisfies the corresponding equality. These can be written by the following equations.

$$\left\{\begin{array}{l} \mathcal{X}(0) = \mathcal{A}^= \otimes \mathcal{X}(0) \oplus \mathcal{A}^+ \otimes \mathcal{X}(1) \oplus \mathcal{X}_0^- \\ \mathcal{X}(k) = \mathcal{A}^= \otimes \mathcal{X}(k) \oplus \mathcal{A}^- \otimes \mathcal{X}(k-1) \oplus \\ \mathcal{A}^+ \otimes \mathcal{X}(k+1) \text{ for } k = 1 \text{ to } h-1 \\ \mathcal{X}(h) = \mathcal{A}^= \otimes \mathcal{X}(h) \oplus \mathcal{A}^- \otimes \mathcal{X}(h-1) \end{array}\right. \qquad [2.15]$$

The following proposition $\mathcal{P}(k)$ is now proved by recursion.

$\mathcal{P}(k)$: $\mathcal{X}^-(k) = (w_k)^* \otimes [\mathcal{A}^+ \otimes \mathcal{X}^-(k+1) \oplus \beta_k^-]$

Base case: $\mathcal{P}(0)$

From the first equality of [2.15], we can write

$\mathcal{X}(0) = w_0 \otimes \mathcal{X}(0) \oplus \mathcal{A}^+ \otimes \mathcal{X}(1) \oplus \beta_0^-$, where $w_0 = \mathcal{A}^=$ and $\beta_0^- = \mathcal{X}_0^-$. Therefore, $\mathcal{X}(0) = (w_0)^*[\mathcal{A}^+ \otimes \mathcal{X}(1) \oplus \beta_0^-]$, which proves $\mathcal{P}(0)$.

Case: $\mathcal{P}(1)$

From the second equality of [2.15], we can write, for $k = 1$

$\mathcal{X}(1) = \mathcal{A}^= \otimes \mathcal{X}(1) \oplus \mathcal{A}^- \otimes \mathcal{X}(0) \oplus \mathcal{A}^+ \otimes \mathcal{X}(2)$

If $\mathcal{P}(0)$ is used,

$\mathcal{X}(1) = \mathcal{A}^{=} \otimes \mathcal{X}(1) \oplus \mathcal{A}^{-} \otimes [(w_0)^*[\mathcal{A}^{+} \otimes \mathcal{X}(1) \oplus \beta_0^-]] \oplus \mathcal{A}^{+} \otimes \mathcal{X}(2)$

The distributivity of $\otimes$ with respect to $\oplus$ leads to

$\mathcal{X}(1) = [\mathcal{A}^{=} \oplus \mathcal{A}^{-} \otimes (w_0)^* \otimes \mathcal{A}^{+}] \otimes \mathcal{X}(1) \oplus \mathcal{A}^{-} \otimes (w_0)^* \otimes \beta_0^- \oplus \mathcal{A}^{+} \otimes \mathcal{X}(2) = w_1 \otimes \mathcal{X}(1) \oplus \beta_1^- \oplus \mathcal{A}^{+} \otimes \mathcal{X}(2)$, where $w_1 = \mathcal{A}^{=} \oplus \mathcal{A}^{-}(w_0)^* \mathcal{A}^{+}$ and $\beta_1^- = \mathcal{A}^{-}(w_0)^* \otimes \beta_0^-$. Therefore,

$\mathcal{X}(1) = (w_1)^* \otimes [\mathcal{A}^{+} \otimes \mathcal{X}(2) \oplus \beta_1^-]$ and $\mathcal{P}(1)$ is proved. Now, this approach is generalized for $k = 1$ to $h - 1$.

Case: $\mathcal{P}(k)$ for k from 1 to $h - 1$.

Let us assume $\mathcal{P}(k-1)$: $\mathcal{X}(k-1) = (w_{k-1})^* \otimes [\mathcal{A}^{+} \otimes \mathcal{X}(k) \oplus \beta_{k-1}^-]$. We will prove that $\mathcal{P}(k-1)$ entails $\mathcal{P}(k)$.

From the second equality of [2.15], we can write

$\mathcal{X}(k) = \mathcal{A}^{=} \otimes \mathcal{X}(k) \oplus \mathcal{A}^{-} \otimes \mathcal{X}(k-1) \oplus \mathcal{A}^{+} \otimes \mathcal{X}(k+1)$

As $\mathcal{X}(k-1) = (w_{k-1})^* \otimes [\mathcal{A}^{+} \otimes \mathcal{X}(k) \oplus \beta_{k-1}^-]$, the following expression is deduced:

$\mathcal{X}(k) = \mathcal{A}^{=} \otimes \mathcal{X}(k) \oplus \mathcal{A}^{-} \otimes (w_{k-1})^* \otimes [\mathcal{A}^{+} \otimes \mathcal{X}(k) \oplus \beta_{k-1}^-] \oplus \mathcal{A}^{+} \otimes \mathcal{X}(k+1)$

The distributivity of $\otimes$ with respect to $\oplus$ yields

$\mathcal{X}(k) = [\mathcal{A}^{=} \oplus \mathcal{A}^{-} \otimes (w_{k-1})^* \otimes \mathcal{A}^{+}] \otimes \mathcal{X}(k) \oplus \mathcal{A}^{-} \otimes (w_{k-1})^* \otimes \beta_{k-1}^- \oplus \mathcal{A}^{+} \otimes \mathcal{X}(k+1) = w_k \otimes \mathcal{X}(k) \oplus \beta_k^- \oplus \mathcal{A}^{+} \otimes \mathcal{X}(k+1)$, where $w_k = \mathcal{A}^{=} \oplus \mathcal{A}^{-} \otimes (w_{k-1})^* \otimes \mathcal{A}^{+}$ and $\beta_k^- = \mathcal{A}^{-} \otimes (w_{k-1})^* \otimes \beta_{k-1}^-$

Therefore, $\mathcal{X}(k) = (w_k)^*[\mathcal{A}^{+} \otimes \mathcal{X}(k+1) \oplus \beta_k^-]$ and the desired expression is obtained: $\mathcal{P}(k)$ has been deduced from $\mathcal{P}(k-1)$. Moreover, as $\mathcal{P}(0)$ is true, $\mathcal{P}(k)$ has been proved for k from 1 to $h - 1$: the recursion is finished. Knowing β_k^-, the calculation of

$\mathcal{X}(k)$ uses a backward iteration, while the calculation of β_k^- is relevant to a forward iteration.

Now, the final case will be proved.

Case: $\mathcal{P}(h)$

The last equality of [2.15] can be considered like the second equality but without $\mathcal{A}^+ \otimes \mathcal{X}(k+1)$: the argument of case $\mathcal{P}(k)$ can be taken and we can write

$\mathcal{X}(h) = (w_h)^* \otimes \beta_h^-$ with $w_h = \mathcal{A}^= \oplus \mathcal{A}^- \otimes (w_{h-1})^* \otimes \mathcal{A}^+$ and $\beta_h^- = \mathcal{A}^- \otimes (w_{h-1})^* \otimes \beta_{h-1}^-$.

Finally, as matrices w_k have no positive circuit and $\mathcal{X}_0^-$ belongs to $(\mathbb{R} \cup \{-\infty\})^n$, the state trajectory is defined in $\mathbb{R} \cup \{-\infty\}$. □

2.6.2. *Greatest state trajectory*

The following theorem can be deduced from the previous theorem by duality. Steps (a) are identical.

THEOREM 2.4.– *If the process operates on horizon $h \in \mathbb{N}$ and if matrices w_k defined below have no positive circuit, the greatest state trajectory in $\mathbb{R} \cup \{+\infty\}$ satisfying $\mathcal{X}(0) \leq \mathcal{X}_0^+ \in (\mathbb{R} \cup \{+\infty\})^n$ is given by the following forward / backward algorithm.*

Forward/backward algorithm

a) Coefficients of w_k by forward iteration

Initialization: $w_0 = \mathcal{A}^=$

for $k = 1$ *to* h, $w_k = \mathcal{A}^= \oplus \mathcal{A}^- \otimes (w_{k-1})^* \otimes \mathcal{A}^+$

b) First estimate β_k^+ by forward iteration

Initialization: $\beta_0^+ = \mathcal{X}_0^+$

for $k = 1$ *to* h, $\beta_k^+ = ((w_{k-1})^* \otimes \mathcal{A}^+) \backslash \beta_{k-1}^+$

c) Trajectory $\mathcal{X}_k^+$ by backward iteration

$\mathcal{X}_h^+ = (w_h)^* \backslash \beta_h^+$

for $k = h - 1$ *to* 0, $\mathcal{X}_k^+ = (w_k)^* \backslash [\mathcal{A}^- \backslash \mathcal{X}_{k+1}^+ \wedge \beta_k^+]$

PROOF.– The proof is omitted as it can be deduced by duality from the previous theorem. □

To sum up, the two algorithms allow the determination of the lowest (respectively, greatest) acceptable trajectories satisfying $\mathcal{X}(0) \geq \mathcal{X}_0^-$ (respectively, $\mathcal{X}(0) \leq \mathcal{X}_0^+$). They also allow the checking of the existence of a trajectory satisfying $\mathcal{X}(0) \in [\mathcal{X}_0^-, \mathcal{X}_0^+]$ if constraints $\mathcal{X}(0) \leq \mathcal{X}_0^+$ and $\mathcal{X}(0) \geq \mathcal{X}_0^-$, respectively, are added in the corresponding algorithms.

REMARK 2.3.– Defined on a box $[\mathcal{X}_0^-, \mathcal{X}_0^+]$, the initial condition is less restrictive than the more usual $\mathcal{X}(0) = \mathcal{X}_0$, which is a particular case. In a natural way, checking this case is made as follows. The determination of the lowest trajectory such as $\mathcal{X}(0) \in [\mathcal{X}_0, \mathcal{X}_0^+]$, with $\mathcal{X}_0 \leq \mathcal{X}_0^+$, allows checking the acceptability of $\mathcal{X}_0$ or in other words, if there is a solution $\mathcal{X}$ so that $\mathcal{X}(0) = \mathcal{X}_0$. Similarly, the determination of the greatest trajectory such as $\mathcal{X}(0) \in [\mathcal{X}_0^-, \mathcal{X}_0]$ with $\mathcal{X}_0^- \leq \mathcal{X}_0$ also allows checking the existence of a solution $\mathcal{X}$ so that $\mathcal{X}(0) = \mathcal{X}_0$.

REMARK 2.4.– The calculation of the state trajectories starts from values $(w_h)^* \otimes \beta_h^-$ and $(w_h)^* \backslash \beta_h^+$ and consequently depends on horizon h. The calculation of w_k, β_k^- and β_0^+ depends on index k, but not on horizon h.

2.7. Consistency

In this chapter, an acceptable behavior of the considered P-time event graph is defined by any operation guaranteeing the liveness of tokens. Therefore, it does not lead to any deadlock situation. As this behavior is represented by the algebraic model [2.12], the aim of this part is to study the existence of a state trajectory solution to these inequalities. Clearly, if we can calculate an arbitrary trajectory starting from box $[\mathcal{X}_0^-, \mathcal{X}_0^+]$, we can deduce that the system is consistent on the considered horizon. We introduce the following notation.

DEFINITION 2.3.– *A dynamic associated graph of square matrices A, B and C on horizon h, denoted by $G_h(A, B, C)$, is deduced from these matrices by associating a node j_k with column j and a node i_k with row i, for $k = 0$ to h. The pattern is as follows: (a) an arc from node j_{k-1} toward node i_k if $A_{ij} \neq \varepsilon$; (b) an arc from node j_k toward node i_k if $B_{ij} \neq \varepsilon$; (c) an arc from node j_k toward node i_{k-1} if $C_{ij} \neq \varepsilon$.*

In this chapter, we consider $G_h(\mathcal{A}^-, \mathcal{A}^=, \mathcal{A}^+)$. An example is given in Figure 2.5. The dimension of each column k is the dimension n of the state. As in the static case presented in section 2.4, the system is consistent if the dynamic associated graph $G_h(\mathcal{A}^-, \mathcal{A}^=, \mathcal{A}^+)$ has no circuit with positive weight. These circuits can simply be situated in the associated graph of the static part ($\mathcal{A}^=$) or of the dynamic part ($\mathcal{A}^-$ and $\mathcal{A}^+$). Figure 2.7 shows that the circuits can present a complex form.

Assuming the liveness of the event graph, the following theorem considers the temporal consistency of P-time event graphs. This theorem concerns the existence of a state trajectory in $\mathbb{R}$, and not in $\mathbb{R}_{\max}$.

THEOREM 2.5.– *A live P-time event graph is consistent on arbitrary horizon h if and only if the dynamic associated graph $G_h(\mathcal{A}^-, \mathcal{A}^=, \mathcal{A}^+)$ contains circuits with only non-positive weight.*

PROOF.– The model can be completely represented by system [2.15] after replacing symbol $=$ with $\geq$. This system can be rewritten in $\mathbb{R}_{\max}$ under the global form $x \geq A \otimes x \oplus B$, which includes every inequality. The relevant dynamic associated graph is $G_h(\mathcal{A}^-, \mathcal{A}^=, \mathcal{A}^+)$. As the least solution is A^*B, this system has at least one solution in $\mathbb{R}_{\max}$ if the global matrix A has strictly no positive circuits. This gives a sufficient condition for the existence of a solution in $\mathbb{R}_{\max}$. Moreover, $\mathcal{X}_0^-$ can be taken finite: as $\mathcal{X}(0) \geq \mathcal{X}_0^-$ and as the trajectory is non-decreasing by construction of $\mathcal{A}^-$, each component of the state trajectory is different from ε and the trajectory belongs to $\mathbb{R}$. Conversely, if finite $\mathcal{X}$ satisfies the model of the P-time event graph, it can only satisfy subsystems with non-positive circuits. □

The following result is immediate.

COROLLARY 2.2.– A live P-time event graph with a null initial marking is consistent if and only if the dynamic associated graph of matrix $\mathcal{A}^=$ contains circuits with only non-positive weight.

REMARK 2.5.– A live P-time event graph whose initial marking is null is without circuit (in the event graph), but the dynamic associated graph of matrix $\mathcal{A}^=$ can have circuits.

Now, we consider matrices w_k that allow a characterization of the circuits. The following property gives a graphical interpretation of the calculation of these matrices.

PROPERTY 2.1.– Entry $((w_h)^*)_{i_h, j_h}$ represents the maximum weight of all the paths from vertices j_h to vertices i_h for $i, j \in [1..n]$ in the dynamic associated graph $G_h(\mathcal{A}^-, \mathcal{A}^=, \mathcal{A}^+)$ except the paths containing an arc from the node with index $j0$ to the node with index $i0$.

PROOF.– Let us consider the relations inside horizon [0, 1]. So, $\mathcal{X}(1)_i \geq (\mathcal{A}^-)_{i,l} \otimes \mathcal{X}(0)_l \geq (\mathcal{A}^-)_{i,l} \otimes (\mathcal{A}^+)_{l,j}\mathcal{X}(1)_j$, and also,

$\mathcal{X}(1)_i \geq (\mathcal{A}^=)_{i,j}\mathcal{X}(1)_j$. So, $(w_1)_{i,j} = (\mathcal{A}^= \oplus \mathcal{A}^- \otimes \mathcal{A}^+)_{i,j}$ represents the greatest weight on the paths composed of the following subpaths:

– an arc $(j_1 \rightarrow i_1)$ (matrix $\mathcal{A}^=$);

– or two successive arcs $(j_1 \rightarrow l_0)$ and $(l_0 \rightarrow i_1)$ (product $\mathcal{A}^- \otimes \mathcal{A}^+$).

Note that $(w_1)^*$ is not the result of paths containing an arc from the node with index j_0 to the node with index i_0 directly, as there is no term like $\mathcal{A}^- \otimes \mathcal{A}^= \otimes \mathcal{A}^+$.

For horizon [0, 2], $(w_2)_{ij} = (\mathcal{A}^= \oplus \mathcal{A}^- \otimes (w_1)^* \otimes \mathcal{A}^+)_{ij}$ represents the greatest weight on the paths composed of the following subpaths:

– an arc $(j_2 \rightarrow i_2)$ (matrix $\mathcal{A}^=$); or

– an arc $(j_2 \rightarrow l_1)$ (matrix $\mathcal{A}^+$), the previous paths from l_1 to m_1 expressed by matrix $(w_1)^*$ (described above) and an arc $(l_1 \rightarrow i_2)$ (matrix $\mathcal{A}^-$).

Consequently, $(w_2)^*$ represents the greatest weight of every path (and circuit) of the associated graph on horizon [0, 2] and defined by a path from i_2 to i_2 and going possibly to nodes of indexes 1 and 0, except the paths containing an arc from nodes of indexes 0 to 0.

The argument can be repeated until $k = h$. □

The following theorem improves Theorem 2.5 by giving a practical way to check the consistency.

THEOREM 2.6.– *A live P-time event graph is consistent if and only if the associated graph of each matrix w_k for any $k \geq 0$ contains circuits with only non-positive weight.*

PROOF.– Property 2.1 says that matrices $(w_k)^*$ represent the greatest weight of almost every path and circuit in

$G_h(\mathcal{A}^-, \mathcal{A}^=, \mathcal{A}^+)$. The weights of circuits that are not "present" in $(w_k)^*$ are "present" in $(w_{k+1})^*$ for $G_{h+1}(\mathcal{A}^-, \mathcal{A}^=, \mathcal{A}^+)$, as the associated graph is the repetition of a pattern. Consequently, each circuit is expressed and the proof can be deduced from Theorem 2.5. □

Therefore, if there is an index k_1 such that an entry $((w_k)^*)_{i,j}$ is infinite, we can conclude that there is a path from j to i, containing a circuit with a positive weight in $G_{k_1}(\mathcal{A}^-, \mathcal{A}^=, \mathcal{A}^+)$. So, the system is not consistent on horizon greater than $h \geq k_1$. An example of a non-obvious circuit with positive circuit is given in Figure 2.7.

The following result will facilitate the analysis of the consistency and its checking.

PROPERTY 2.2.– $w_k \geq w_{k-1}$ for $k \geq 1$.

PROOF.– Let us suppose that $w_{k-1} \geq w_{k-2}$, $w_k = \mathcal{A}^= \oplus \mathcal{A}^- \otimes (w_{k-1})^* \otimes \mathcal{A}^+ = \mathcal{A}^= \oplus \mathcal{A}^- \otimes \mathcal{A}^+ \oplus \mathcal{A}^- \otimes (w_{k-1}) \otimes \mathcal{A}^+ \oplus \mathcal{A}^- \otimes (w_{k-1})^2 \otimes \mathcal{A}^+ \oplus \ldots$

As $w_{k-1} \geq w_{k-2}$ and by isotony of the product, $w_k \geq \mathcal{A}^= \oplus \mathcal{A}^- \otimes \mathcal{A}^+ \oplus \mathcal{A}^- \otimes (w_{k-2}) \otimes \mathcal{A}^+ \oplus \mathcal{A}^- \otimes (w_{k-2})^2 \otimes \mathcal{A}^+ \oplus \ldots = w_{k-1}$

Moreover, $w_1 \geq w_0 = \mathcal{A}^=$ as $w_1 = \mathcal{A}^= \oplus \mathcal{A}^- \otimes (w_0)^* \otimes \mathcal{A}^+$. So, the series $w_0 = \mathcal{A}^=$ and $w_k = \mathcal{A}^= \oplus \mathcal{A}^- \otimes (w_{k-1})^* \otimes \mathcal{A}^+$ for $k \geq 1$ is non-decreasing. □

Suppose that there is an index k_1 such that matrix w_k does not increase ($w_{k_1+1} = w_{k_1}$) with w_k belonging to $\mathbb{R}_{\max}$. From Property 2.2, we can conclude that no matrix w_k has positive circuit for any index k. Consequently, the P-time event graph is consistent on an infinite horizon. In this case, the tests show that the convergence of consistent P-time event graphs is fast.

2.7.1. *Example 3*

The following example illustrates the results concerning consistency and extremal trajectories. Computation tests are made using the max-plus toolbox in Scilab.

2.7.1.1. *Model*

A slight modification of example 2 in Figure 2.3 gives the closed-loop structure of example 1 in Figure 2.2, which describes a limitation of resources: an upper bound on packing (T_4^+) is added. The initial marking is $(1 \quad 1 \quad 1 \quad 1)^t$ and $T_1^- = 3$, $T_2^- = 3$, $T_3^- = 1$, $T_4^- = 3$, $T_3^+ = 2$, $T_4^+ = 11$. Example 3 is almost similar to example 1 but with a modification of T_2^-. Therefore,

$$A_0^- = \begin{pmatrix} \varepsilon & \varepsilon & \varepsilon \\ \varepsilon & \varepsilon & \varepsilon \\ \varepsilon & \varepsilon & \varepsilon \end{pmatrix} A_0^+ = \begin{pmatrix} \varepsilon & \varepsilon & \varepsilon \\ \varepsilon & \varepsilon & \varepsilon \\ \varepsilon & \varepsilon & \varepsilon \end{pmatrix}$$

$$\mathbb{A}_0^= = A_0^- \oplus A_0^+ = \varepsilon$$

$$A_1^- = \begin{pmatrix} \varepsilon & \varepsilon & 3 \\ 3 & \varepsilon & \varepsilon \\ 1 & 3 & \varepsilon \end{pmatrix} A_1^+ = \begin{pmatrix} \varepsilon & \varepsilon & -2 \\ \varepsilon & \varepsilon & \varepsilon \\ -11 & \varepsilon & \varepsilon \end{pmatrix}$$

Matrices $\mathcal{A}^=$, $\mathcal{A}^-$ and $\mathcal{A}^+$ are now deduced.

$\mathcal{A}^= = \mathbb{A}_0^= = A_0^- \oplus A_0^+ = \varepsilon$, $\mathcal{A}^- = \mathbb{A}_1^- = (\mathbb{A}_0^=)^* \otimes A_1^- = A_1^-$ and $\mathcal{A}^+ = \mathbb{A}_1^+ = (\mathbb{A}_0^=)^* \otimes A_1^+ = A_1^+$. The relevant associated graph is shown in Figure 2.5.

2.7.1.2. *Series*

$w_0 = \mathcal{A}^= = \varepsilon$

$$w_1 = \begin{pmatrix} -8 & \varepsilon & \varepsilon \\ \varepsilon & \varepsilon & 1 \\ \varepsilon & \varepsilon & -1 \end{pmatrix}, w_2 = \begin{pmatrix} -8 & \varepsilon & \varepsilon \\ \varepsilon & \varepsilon & 1 \\ -7 & \varepsilon & -1 \end{pmatrix}, w_3 = \begin{pmatrix} -8 & \varepsilon & -6 \\ \varepsilon & \varepsilon & 1 \\ -7 & \varepsilon & -1 \end{pmatrix}, w_4 = \begin{pmatrix} -8 & \varepsilon & -6 \\ -14 & \varepsilon & 1 \\ -7 & \varepsilon & -1 \end{pmatrix}$$

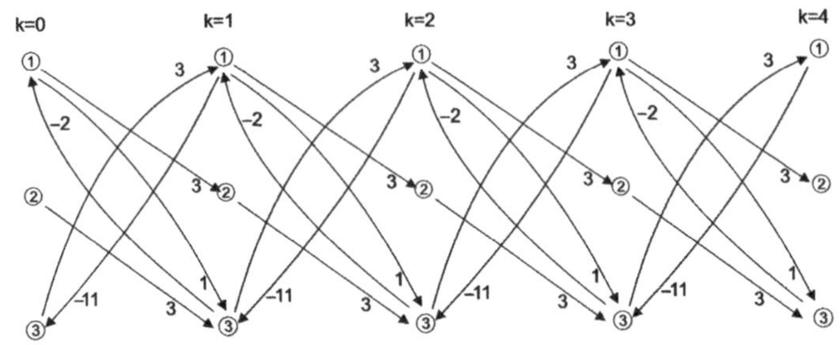

Figure 2.5. *Associated graph in consistent case*

The calculation of matrices w_k shows that they are constant after a short transitory period ($w_k = w_4$ for $k \geq 5$) and that they have no positive circuit. Consequently, the system is consistent on an arbitrary horizon.

2.7.1.3. *Lowest and greatest state trajectories*

Now, we apply the algorithms of Theorems 2.3 and 2.4, which provide the lowest and greatest state trajectories, χ^+ and χ^+, respectively. The arbitrary initial conditions are $\chi_0^- = (1\ \ 0\ \ 0)^t$ and $\chi_0^+ = (1\ \ 0\ \ 0)^t$. In step (b), intermediate values β^- and β^+, which are given in the following tables, are deduced from matrices w_k and initial conditions χ_0^- and χ_0^+ by a forward approach. They give a first estimate of the lowest and greatest trajectories.

k	0	1	2	3	4	5	6	7	8	9	10
β_1^-	1	3	6	10	14	18	22	26	30	34	38
β_2^-	0	4	6	9	13	17	21	25	29	33	37
β_3^-	0	3	7	11	15	19	23	27	31	35	39

k	0	1	2	3	4	5	6	7	8	9	10
β_1^+	1	11	14	24	27	36	40	49	53	62	66
β_2^+	0	T	T	T	T	T	T	T	T	T	T
β_3^+	0	3	13	16	25	29	38	42	51	55	64

Intermediate values β^- and β^+ are now improved by the backward step (c). The following tables contain the lowest and greatest state trajectories, χ^- and χ^+, respectively.

k	0	1	2	3	4	5	6	7	8	9	10
$\mathcal{X}_1^-$	1	5	9	13	17	21	25	29	33	37	38
$\mathcal{X}_2^-$	0	4	8	12	16	20	24	28	32	36	40
$\mathcal{X}_3^-$	0	3	7	11	15	19	23	27	31	35	39

k	0	1	2	3	4	5	6	7	8	9	10
$\mathcal{X}_1^+$	1	10	14	23	27	36	40	49	53	62	66
$\mathcal{X}_2^+$	0	9	13	22	26	35	39	48	52	61	T
$\mathcal{X}_3^+$	0	3	12	16	25	29	38	42	51	55	64

The following table is the state trajectory of the timed event graph using the lower bound of the temporizations of the P-time event graph (see Figure 2.2). With the assumption of earliest behavior, $x(k) = A \otimes x(k)$ with $A = A_1^-$. As $x(0) = \chi_0^- = \begin{pmatrix} 1 & 0 & 0 \end{pmatrix}^t$, a comparison can be made.

k	0	1	2	3	4	5	6	7	8	9	10
x_1	1	3	6	10	12	15	19	21	24	28	30
x_2	0	4	6	9	13	15	18	22	24	27	31
x_3	0	3	7	9	12	16	18	21	25	27	30

Figure 2.6 shows the trajectories of transition x_1: the lowest and greatest trajectories for the P-time event graph; the trajectory of the relevant timed event graph. The three trajectories are clearly different, $(x(k) \leq \chi^-(k) \leq \chi^+(k))$ as their rates.

REMARK 2.6.– It is important to note that each extremal trajectory depends on the lower and upper bounds of the temporizations and not just on one of them. The calculation of the minimal trajectory naturally requires not only the inequalities corresponding to timed event graphs [2.5] but also the upper constraints [2.6]: the example in Figure 2.6 shows that the earliest functioning of a timed event graph using the relevant equality of [2.5] does not satisfy the inequalities of the P-time event graph in the autonomous case. This fact entails that the trajectories of P-time event graphs cannot always be deduced by a direct forward iteration like in the state equation in timed event graphs. Note that in this example, the P-time event graph is consistent.

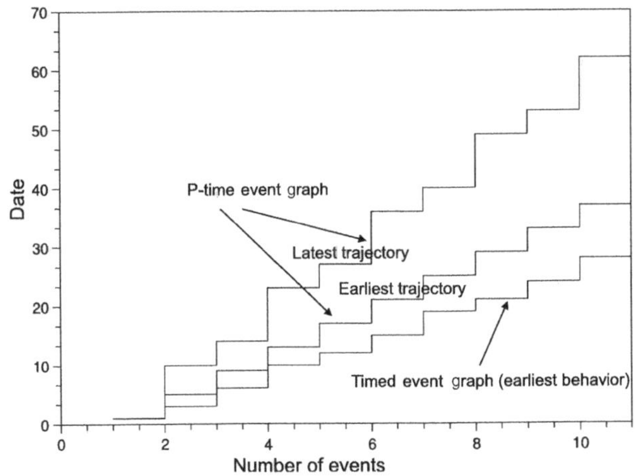

Figure 2.6. *Trajectories*

2.7.2. *Maximal horizon of temporal consistency*

Assuming the liveness of the event graph, we consider the temporal consistency of P-time event graphs. Clearly, if we can calculate an arbitrary finite trajectory (e.g. in $\mathbb{R}$) starting from $\mathcal{X}(0) \in (\mathbb{R})^n$, the system is consistent on the given horizon. Therefore, the liveness of tokens is guaranteed and it does not lead to any deadlock situation. In the previous part, we have proven that the existence of a finite trajectory only depends on matrices w_k and, more precisely, that a finite trajectory exists if and only if matrices $(w_k)^*$ converge in $\mathbb{R}_{\max}$.

Let us now consider the problem of the determination of the maximal horizon of temporal consistency. In step (c) of the algorithms, the calculations of the state trajectory $\mathcal{X}^-(k)$ start from values w_h, β_h^- and β_h^+ and consequently depend on horizon $[0, h]$ where h is a datum. Contrary to step (c), the calculations of w_k, β_k^- and β_k^+ start from $\mathcal{A}^=$, $\mathcal{X}_0^-$ and $\mathcal{X}_0^+$ in steps (a) and (b): they depend on index k, but not on horizon h as the calculations can continue after h. Therefore, the problem now is to determine the maximal horizon $h_{\max}$ where the system can follow a finite trajectory. As the horizon can be finite or infinite, we consider the two following cases:

– Case 1. Matrix $(w_k)^*$ does not belong to $\mathbb{R}_{\max}$. As there is at least one infinite entry $((w_k)^*)_{i,j} = +\infty$, the system is not consistent on horizon $[0, h]$ with $h \geq k$.

– Case 2. Matrix $(w_k)^*$ belongs to $\mathbb{R}_{\max}$. If $w_k = w_{k-1}$, then the P-time event graph is consistent on an infinite horizon and $h_{max} = +\infty$.

A practical way to determine the horizon of consistency h_{max} is as follows.

ALGORITHM.– Initialization: $k \leftarrow 0$. Calculate and analyze $(w_k)^*$ for $k \geq 0$. Stop if case 1 ($h_{\max} = k - 1$ if $k \geq 1$) or case 2 ($h_{\max} = +\infty$) defined above is satisfied or repeat with $k \leftarrow k+1$. □

As the series w_k is non-decreasing, each entry converges to a stable finite value or the infinite value $+\infty$.

Example 1 (section 2.3.1) continued. Now, assume that a failure appears in moving the packet whose normal duration T_2^- associated with p_2 is equal to 3 in example 3, which corresponds to $(\mathcal{A}_1^-)_{3,2}$: the duration is now equal to 6, which corresponds to example 1. So, $\mathcal{A}^{=} = \varepsilon$, $\mathcal{A}^{-} = \begin{pmatrix} \varepsilon & \varepsilon & 3 \\ 3 & \varepsilon & \varepsilon \\ 1 & 6 & \varepsilon \end{pmatrix}$ and $\mathcal{A}^{+} = \begin{pmatrix} \varepsilon & \varepsilon & -2 \\ \varepsilon & \varepsilon & \varepsilon \\ -11 & \varepsilon & \varepsilon \end{pmatrix}$. The relevant matrices w_k are as follows. We also give w_3^*.

$$w_0 = \mathcal{A}^{=} = \varepsilon,\ w_1 = \begin{pmatrix} -8 & \varepsilon & \varepsilon \\ \varepsilon & \varepsilon & 1 \\ \varepsilon & \varepsilon & -1 \end{pmatrix},\ w_2 = \begin{pmatrix} -8 & \varepsilon & \varepsilon \\ \varepsilon & \varepsilon & 1 \\ -4 & \varepsilon & -1 \end{pmatrix},$$

$w_3 = \begin{pmatrix} -8 & \varepsilon & -3 \\ \varepsilon & \varepsilon & 1 \\ -4 & \varepsilon & 1 \end{pmatrix}$, $w_3^* = \begin{pmatrix} T & \varepsilon & T \\ T & 0 & T \\ T & \varepsilon & T \end{pmatrix}$ and $w_4 = \begin{pmatrix} T & \varepsilon & T \\ T & \varepsilon & T \\ T & \varepsilon & T \end{pmatrix}$.

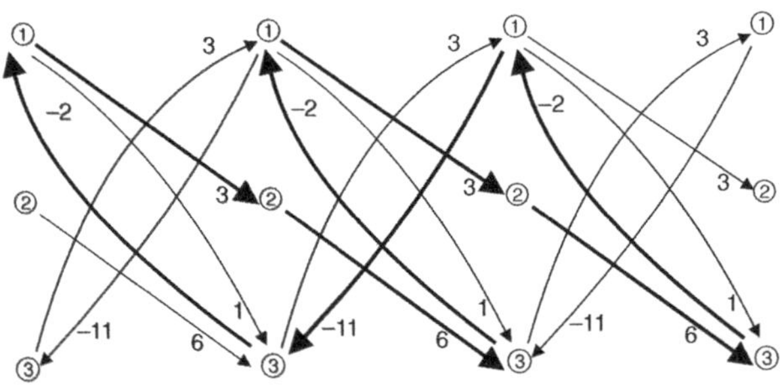

Figure 2.7. *Circuit with positive weight in inconsistent case*

As some coefficients of w_3^* and also of w_4 are equal to $T = +\infty$, the system is not consistent. Therefore, even if the underlying graph of the event graph is live in the usual sense, it is not consistent for $T_2^- = 6$. So, as $(w_3)_{3,3} = 1$, $\chi_3^-(k) \geq 1 \otimes \chi_3^-(k)$, which is inconsistent. This incoherence comes from the following inequalities. Figure 2.7 shows the relevant circuit with positive weight.

$$\chi_3^-(k) \geq 6 \otimes \chi_2^-(k-1)$$

$$\chi_2^-(k-1) \geq 3 \otimes \chi_1^-(k-2)$$

$$\chi_1^-(k-2) \geq (-2) \otimes \chi_3^-(k-1)$$

$$\chi_3^-(k-1) \geq 6 \otimes \chi_2^-(k-2)$$

$$\chi_2^-(k-2) \geq 3 \otimes \chi_1^-(k-3)$$

$$\chi_1^-(k-3) \geq (-2) \otimes \chi_3^-(k-2)$$

$$\chi_3^-(k-2) \geq (-11) \otimes \chi_1^-(k-1)$$

$$\chi_1^-(k-1) \geq (-2) \otimes \chi_3^-(k)$$

As some coefficients of w_3^* are equal to $+\infty$, $h_{\max} = 2$ and a trajectory can only be defined on horizon $[0, 2]$. System [2.6] is only consistent for $k = 0$ and 1. The trials show that the tolerance margin of T_2^-, where the system is consistent, is $[0, 5.5]$. □

2.7.3. *Date of the first token deaths*

If the system is only consistent on horizon $h_{\max}$, an admissible trajectory can be calculated but the tokens produced by the firing at date $\mathcal{X}(h_{\max})$ do not lead to a complete firing of the transitions at the following number of events $h_{\max} + 1$. In the following text, we consider only the case of places with unitary initial marking: the case of places with a null initial marking is omitted but follows a similar technique. If $m_l = 1$, the time interval of token stay is $[\mathcal{A}_{ig}^- \otimes \mathcal{X}_g(k), \mathcal{A}_{gi}^+ \backslash \mathcal{X}_g(k)]$ for a token generated by the kth firing of transition g in a place $p_l \in P$ such that $\mathcal{X}_g \in^\bullet p_l$ and $\mathcal{X}_i \in p_l^\bullet$. As there is at least one transition i such that relation $\bigoplus_{j\in^\bullet(^\bullet\mathcal{X}_i)} \mathcal{A}_{ij}^- \otimes \mathcal{X}_j(h_{\max}) \leq \mathcal{X}_i(h_{\max}+1) \leq \bigwedge_{j\in^\bullet(^\bullet\mathcal{X}_i)} \mathcal{A}_{ji}^+ \backslash \mathcal{X}_j(h_{\max})$ is not satisfied, the non-synchronization of transition i leads to some token deaths. Let ST be the set of transitions $g \in^\bullet (^\bullet\mathcal{X}_i)$

such that $\mathcal{A}^+_{gi} \backslash \mathcal{X}_j(h_{\max}) = \bigwedge_{j \in {}^\bullet({}^\bullet \mathcal{X}_i)} \mathcal{A}^+_{ji} \backslash \mathcal{X}_j(h_{\max})$. Each input transition $g \in ST$ generates a token, which dies in the place $p_l \in (\mathcal{X}_g)^\bullet \bigcap {}^\bullet(\mathcal{X}_i)$ at the date $\mathcal{A}^+_{gi} \backslash \mathcal{X}_j(h_{\max})$.

However, the firing of transition i is still possible if a new firing of each transition $g \in ST$ produces another token. This can be expressed by a shift in the numbering of the events. Therefore, relation $\mathcal{A}^-_{ig} \otimes \mathcal{X}_g(k) \leq \mathcal{X}_i(k+1) \leq \mathcal{A}^+_{gi} \backslash \mathcal{X}_g(k)$ for $k < h_{\max}$ becomes relation $\mathcal{A}^-_{ig} \otimes \mathcal{X}_g(k+1) \leq \mathcal{X}_i(k+1) \leq \mathcal{A}^+_{gi} \backslash \mathcal{X}_g(k+1)$ for $k \geq h_{\max}$ in the new algebraic model.

Example 1 (section 2.3.1) continued. For $\mathcal{X}_0^- = \begin{pmatrix} 1 & 0 & 0 \end{pmatrix}^t$, Theorem 2.3 provides the lowest trajectory $\mathcal{X}^-$, which is also trajectory $\mathcal{X}$ given in the first table of the example. Using these dates, we can deduce that the date of the first token death is 13. The new model is as follows: for $k \geq 2$, matrices $\mathcal{B}^=$, $\mathcal{B}^-$ and $\mathcal{B}^+$, replace the previous one in system [2.13].

$$\mathcal{B}^= = \begin{pmatrix} \varepsilon & \varepsilon & -2 \\ \varepsilon & \varepsilon & \varepsilon \\ 1 & \varepsilon & \varepsilon \end{pmatrix}, \mathcal{B}^- = \begin{pmatrix} \varepsilon & \varepsilon & 3 \\ 3 & \varepsilon & \varepsilon \\ \varepsilon & 6 & \varepsilon \end{pmatrix} \text{ and}$$

$$\mathcal{B}^+ = \begin{pmatrix} \varepsilon & \varepsilon & \varepsilon \\ \varepsilon & \varepsilon & \varepsilon \\ -11 & \varepsilon & \varepsilon \end{pmatrix}. \qquad \square$$

2.7.4. *Computational complexity*

The following curve gives indications of the possible CPU (Central Processing Unit) times needed to compute the different matrices w_k, and the lowest and greatest trajectories on an ordinary *Pentium* 1.3 GHz for a horizon $h = 100$. Computation tests are made using maxplus toolboxes under Scilab. The matrices $\mathcal{A}^-$ and $\mathcal{A}^+$ are completely full: there is a place containing a token between each couple of transitions.

For instance, if $n = 50$, the relevant PNs contains 50 transitions and 2,500 places. The matrix $\mathcal{A}^-$ is generated randomly and $\mathcal{A}^+$ is deduced from $\mathcal{A}^-$ such that the system is temporally live on the desired horizon: therefore, the complete calculations are made. In that objective, we also take $\mathcal{A}^= = \varepsilon$, which does not effect the time significantly.

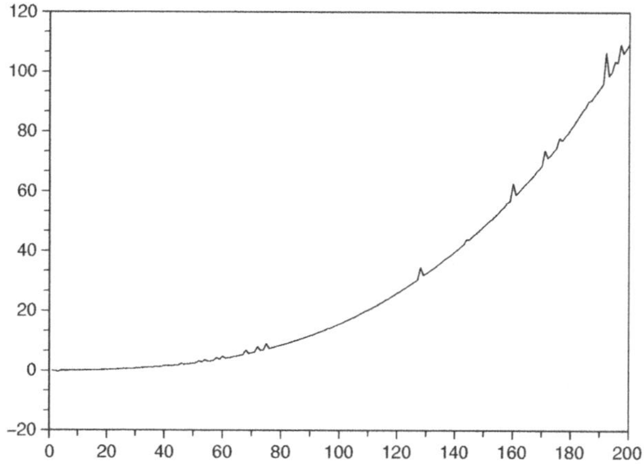

Figure 2.8. *CPU time for different dimensions from 3 to 200 and h = 100*

The algorithms use elementary operations on matrices like $\otimes$, $\oplus$, $\backslash$, $\wedge$ and the more complex operation Kleene star $(*)$. The last operation determines the computational complexity of each step and the complexities of the different known algorithms are polynomial. Therefore, the complexity of calculation of the greatest trajectory is approximately $O(h.n^3)$ with h the horizon and n the dimension of the matrices if the complexity of the used algorithm of Kleene star is $O(n^3)$ (more details can be found in [GON 84]). The space needed for the matrices w_k is $l.n^2$ with l the minimum of the horizon h and the length of the transient period. In short, the algorithm can consider important sizes of event graphs and horizon of calculation.

2.8. Conclusion

Considering the (max, +) model of P-time event graphs, our first objective is the determination of the extremal state trajectories satisfying an initial condition defined on an interval. A direct resolution needs the storage of a matrix whose dimension is $(h.n \times h.n)$ and must follow the different forward/backward relations. On the basis of a specific series of matrices, the proposed resolution is composed of three steps: using the Kleene star, the iterative calculation determines the values of the greatest paths for different horizons; a forward iteration generates a box containing the extremal trajectories and a backward iteration gives the extremal trajectories. The introduction of a non-decreasing series of matrices alleviates the storage as the dimension is the size of the model, which depends on the number of transitions and the initial marking. Therefore, each calculation processes reduced matrices of dimension $(n \times n)$. The approach can be applied to important systems for large horizons because the algorithms are strongly polynomial: the complexity is $O(h.n^3)$ for a given horizon h if the complexity of the used algorithm of Kleene star is $O(n^3)$. Like [GON 84], we give the CPU time for different dimensions and horizons. Moreover, the results obtained in step (a) for a given horizon can be reused in the calculations for a new horizon. The same remark can be made for step (b) if the initial starting point is identical.

The determination of the maximal horizon of temporal consistency is the second objective. The technique is based on the analysis of convergence of matrices w_k^*: each entry can converge to a stable finite value or the infinite value $+\infty$. For a given P-time event graph, the case of a convergence to a constant matrix after a transitory period $h_{\max}$ facilitates the storage and reuse in the calculation of a new trajectory for any horizon. If the system is only consistent on horizon $h_{\max}$, a non-synchronization cannot be avoided at $h_{\max} + 1$ and we calculate the date of the first token deaths.

3

Cycle Time

In this chapter, we enlarge the class of P-time event graphs by introducing interdependent residence durations. This new model is studied through a performance analysis. Considering a periodic behavior, we give conditions of existence of a trajectory and propose different techniques allowing the determination of extremal solutions. The cycle time of the generalized P-time event graph depends not only on the circuits of an associated graph but also on more complex structures. We consider the problem of the maximization/minimization of the cycle time and of the residence duration of some tasks.

3.1. Objectives

In this chapter, we propose a generalization of the P-time event graphs by introducing links between residence durations. Indeed, in some practical examples, some tasks must compensate for the undesirable effects of other operations like the warming of a part or an incomplete achievement. The treatment of this new Petri net needs to be modeled and we will describe its algebraic model in the standard algebra. We will show that the relevant matrices generalize the incidence matrices obtained in previous works [DEC 08].

The second objective is the performance analysis. Classically, the cycle time (or equivalently, the production rate) is based on the consideration of the circuits and determined by the calculation of the maximum of the ratios defined by the sum of temporizations to the sum of the number of the initial tokens, for each elementary circuit [RAM 80]. However, this technique cannot be applied to our new model as the considered matrices are not incidence matrices and cannot be associated with a graph. The well-known theorem [RAM 80] based on the elementary circuits of the associated graph and the algorithm of Karp cannot be applied. Therefore, we propose to use linear programming to solve this new problem, which includes the initial one. We assume that the trajectory follows a 1-periodic behavior, which facilitates the detection of the perturbations by the supervisory control.

A third objective is the optimization of the cycle time and of the residence duration of some tasks. Indeed, we focus on a periodic trajectory leading to an optimal behavior of the tasks where some residence durations are minimized and/or maximized. Practical examples are the minimization of the duration of a task, which leads to energy savings, the minimization of the working time of the employees that have to be paid, or the maximization of the use or the availability of a machine.

To our knowledge, the papers considering cycle time or production rate with linear programming [MAG 84, BUR 91, CAM 92, GIU 00, MUD 05] do not use the theorems of Farkas and Stiemke. We also consider in this chapter the optimization of a general criterion that includes the cycle time but is not limited to it. Moreover, the considered Petri net is original as it is more general than the timed event graph and the P-time event graph. Similar but different to the algebraic model [LIB 96] based on numbers of transition firings ("counter" approach), our algebraic model focuses on the dates of transition firings and a first version has been introduced in [DEC 07b]. Note that this algebraic model can also be used in control [GUE 09].

The chapter is organized as follows. Section 3.2.1 defines the generalized P-time event graphs and its algebraic model. Using a variant of Farkas' lemma, section 3.2.4 gives conditions of consistency of the cycle time. Then, we propose two techniques of linear programming allowing the determination of the cycle times and the dates of transitions' firing. A simple example illustrates the different concepts. The reader can find in the Appendix (section 3.5) the technique allowing the reduction of the horizon associated with the model.

3.2. Problem without optimization

3.2.1. *Objective*

Remember that m is the maximum number of initial tokens: $m = max\{(M_0)_l \mid l \in [1, \mid P \mid]\}$. The objective is now the analysis of a 1-periodic trajectory defined by $x(k+1) = \lambda \times un + x(k)$ where λ is the cycle time and $un = (1, 1, ..., 1)^t$, using the following algebraic model:

$$\begin{pmatrix} Q^- \\ Q^+ \end{pmatrix} \times \begin{pmatrix} x(k-m) \\ x(k-m+1) \\ \vdots \\ x(k-1) \\ x(k) \end{pmatrix} \leq \begin{pmatrix} -T^- \\ T^+ \end{pmatrix} \qquad [3.1]$$

where the dimension of $Q^- = [Q_m^- Q_{m-1}^- Q_{m-2}^- \cdots Q_1^- Q_0^-]$ and $Q^+ = [Q_m^+ Q_{m-1}^+ Q_{m-2}^+ \cdots Q_1^+ Q_0^+]$ is equal to $\mid P \mid . (m+1). \mid TR \mid$.

Without reduction of generality, we suppose that $m = 1$ (see the Appendix (section 3.5)). In Chapter 2, we give the algebraic model of a P-temporal event graph in the (max, +) algebra.

We will show that a more general model (P-time event graph with interdependent residence durations) can be expressed with the previous form expressed in the standard algebra. The existence of a solution and the optimization of a specific criterion are described in the following sections.

3.2.2. *Matrix expression of a P-time event graph*

Let us remember the firing interval for each transition of P-time event graphs, guaranteeing the absence of token-dead states (more details are given in Chapter 2). Note that an event graph can be considered as a set of connected subgraphs made up of a place p_l linked with one upstream transition $\{x_j\} = ^{\bullet} p$ and one downstream transition $\{x_i\} = p^{\bullet}$.

Using the lower bound T_l^- and the upper bound T_l^+, we can write the following system for each place p_l where $(j, i) = (^{\bullet}p, p^{\bullet})$:

$$\begin{cases} x_j(k - (M_0)_l) - x_i(k) \leq -T_l^- \\ -x_j(k - (M_0)_l) + x_i(k) \leq T_l^+ \end{cases}$$

– In the first row (respectively, second row), the weight $+1$ (respectively, -1) of $x_j(k - (M_0)_l)$ is the weight of the arc going from x_j to place p_l, which is equal to $+W_{lj}^+$ (respectively, $-W_{lj}^+$).

– In the first row (respectively, second row), the weight -1 (respectively, $+1$) of $x_i(k)$ is the weight of the arc going from place p_l to transition x_i, which is equal to $-W_{li}^-$ (respectively, $+W_{li}^-$).

The set of previous inequalities that describes a P-time event graph can be expressed with the previous symmetrical form [3.1]: column vectors $-T^-$ and T^+ are vectors of temporization where $[T_l^-, T_l^+]$ is the time interval of place p_l. Each place corresponds to a row of Q^- and a row of Q^+ that contains the weights of its in-going and out-going arcs, which are usually expressed by W^+ and W^-.

– The entry (l, j) of the matrix Q_r^- for $r \in [0, m]$ contains the weight with the plus sign, of the arc going from the transition x_j to the place p_l with an initial marking $(M_0)_l = r$.

– Moreover, the entry (l, j) of the matrix Q_0^- contains the weight with the minus sign, of the arc from the place p_l to the transition x_j, for any initial marking $(M_0)_l \in [0, m]$.

So, $Q_r^- \geq 0$ for $r \in [1, m]$ and a coefficient of the matrix Q_0^- can be null, negative or positive: $Q_0^- \geq -W^-$. The interpretation of the matrix Q^+ is identical to Q^- but with a change of sign of the coefficients: $Q^+ = -Q^-$.

The notation $M_{i,.}$ corresponds to row i of the matrix M. From the above description on the weight of the arcs, we can deduce the following relation expressing the incidence matrix W:

$$W_{l,.} = \sum_{r=0}^{m} (Q_r^-)_{l,.} = -\sum_{r=0}^{m} (Q_r^+)_{l,.} \qquad [3.2]$$

for each place l of the P-temporal event graph.

– Note that $Q_1^- = W^+$ and $Q_0^- = -W^-$ if the initial marking of each place is unitary.

– Note that $Q_0^- = W$ and there is no Q_1^-, if the initial marking is null ($M_0 = 0$).

3.2.3. *Matrix expression of P-time event graphs with interdependent residence durations*

We can also consider the case of residence duration of a token in place p_l, which determines the temporization of another place. For instance, a product that has been cooked cannot be put into a package immediately and needs to cool down. In the food industry, specific cooling systems are used. We can suppose the temporization of the cooling of a part is an affine relation depending on its cooking time. These residence durations are called interdependent.

We assume that the initial marking is null: a realistic assumption is that no part stays in the oven and the cooling system initially. Naturally, the following relations can be rewritten with a non-null initial marking. Using lower bound T_l^- we can write the following inequality for a place p_l where $(x_j, x_i) = ({}^\bullet p_l, p_l^\bullet)$:

$$T_l^- + x_j(k) \leq x_i(k)$$

Let us consider another place p_h where $(x_{j'}, x_{i'}) = ({}^\bullet p_h, p_h^\bullet)$: the residence duration of a token is $x_{i'}(k) - x_{j'}(k)$.

Now, we say that the place p_l is a dependant from the place p_h with an affine function if we can write $T_l^- = a.(x_{i'}(k) - x_{j'}(k)) + b$. Note that b is coherent with a temporization.

Therefore, the inequality obtained for place p_l is $a.(x_{i'}(k) - x_{j'}(k)) + x_j(k) - x_i(k) \leq -b$, which is a relation between four variables. This relation corresponds to a row in Q_0^- with the entries a, $-a$, 1 and -1; the relevant entry in $-T^-$ is $-b$; symmetrically, the entries in Q_0^+ and T^+ are null. When the outstreaming transition $x_{i'}$ of the place p_h is the instreaming transition x_j of the place p_l, the relation becomes a relation between three variables. We have $(1 + a).x_{i'}(k) - a.x_{j'}(k) - x_i(k) \leq -b$.

Containing more variables, these relations are more complex than the relations of the P-time event graphs in their classic form. The coefficients depend on the entries of the incidence matrix but also on the coefficients of the affine function.

3.2.4. *General form* $Ax \leq b$

We now start the resolution of the problem and deduce from [3.1] the following system:

$$\begin{pmatrix} Q_1^- & Q_0^- \\ Q_1^+ & Q_0^+ \end{pmatrix} \cdot \begin{pmatrix} x(k) \\ \lambda \cdot un + x(k) \end{pmatrix} \leq \begin{pmatrix} -T^- \\ T^+ \end{pmatrix}$$

where

$$\begin{pmatrix} (Q_1^- + Q_0^-) \\ (Q_1^+ + Q_0^+) \end{pmatrix} \cdot x(k) + \begin{pmatrix} Q_0^- \\ Q_0^+ \end{pmatrix} \cdot \lambda \cdot un \leq \begin{pmatrix} -T^- \\ T^+ \end{pmatrix} \qquad [3.3]$$

If we simplify the writing with $x(k) = x$, we obtain the following system that presents the general form $Ax \leq b$,

$$\begin{pmatrix} A_{11} & A_{12} \\ A_{21} & A_{22} \end{pmatrix} \cdot \begin{pmatrix} x \\ \lambda \end{pmatrix} \leq \begin{pmatrix} -T^- \\ T^+ \end{pmatrix} \qquad [3.4]$$

where $A_{11} = Q_1^- + Q_0^-$, $A_{12} = Q_0^-.un$, $A_{21} = Q_1^+ + Q_0^+$ and $A_{22} = Q_0^+.un$.

Let us consider the important particular case of a P-temporal event graph. The following property makes the connection with the incidence matrices.

PROPERTY 3.1.– For a P-time event graph, we have $A_{11} = W = -A_{21}$, $A_{12} = -M_0 = -A_{22}$ and system [3.4] becomes:

$$\begin{pmatrix} W & -M_0 \\ -W & M_0 \end{pmatrix} \cdot \begin{pmatrix} x \\ \lambda \end{pmatrix} \leq \begin{pmatrix} -T^- \\ T^+ \end{pmatrix} \qquad [3.5]$$

PROOF.– This can be rewritten because we deduce from [3.2] $Q_1^- + Q_0^- = W$ and $Q_1^+ + Q_0^+ = -W$. So, we can deduce that $A_{11} = W$, $A_{21} = -W$.

Let us note that $Q_0^- = -W^-$ and $Q_1^- = W^+$ when the P-time event graph has one token by place initially.

Thus, $Q_0^- \cdot un = -W^- \cdot un = -M_0$, which represents the initial marking. This result is also true when $m = 1$, that is each place initially has one token at the most. Indeed, each place without a token is represented by a line of Q_0^-, which contains the weights 1 and -1 corresponding to its in-going and outgoing arcs. A similar reasoning holds for the lower part of the system and we obtain $Q_0^+ \cdot un = W^- \cdot un = M_0$. □

3.2.5. *Example*

Model

The state is $x(k) = (x_1(k) \;\; x_2(k) \;\; x_3(k) \;\; x_4(k))^t$. The temporal intervals are: $[T_1^-, T_1^+] = [3, 10]$, $[T_2^-, T_2^+] = [3, 20]$, $[T_3^-, T_3^+] = [1, 2]$, $[T_5^-, +\infty] = [11.5, +\infty]$ and $[T_6^-, T_6^+] = [1, 5]$. The place p_4 describes the cooling down of the product that has been cooked in an oven (place p_2): $T_4 = a.(x_3(k) - x_2(k-1)) + b$ with $a = 5$ and $b = 3$. The matrices of the algebraic model are as follows:

$$T^- = (3 \;\; 3 \;\; 1 \;\; b \;\; 11.5 \;\; 1)^t \text{ and } T^+ = (10 \;\; 20 \;\; 2 \;\; +\infty \;\; +\infty \;\; 5)^t$$

$$Q_1^- = \begin{pmatrix} 1 & 0 & 0 & 0 \\ 0 & 1 & 0 & 0 \\ 1 & 0 & 0 & 0 \\ 0 & -a & 1 & 0 \\ 0 & 0 & 1 & 0 \\ 0 & 1 & 0 & 0 \end{pmatrix}, \; Q_0^- = \begin{pmatrix} 0 & -1 & 0 & 0 \\ 0 & 0 & -1 & 0 \\ 0 & 0 & -1 & 0 \\ -1 & 0 & a & 0 \\ 0 & 0 & 0 & -1 \\ 0 & 0 & 0 & -1 \end{pmatrix},$$

$$Q_1^+ = \begin{pmatrix} -1 & 0 & 0 & 0 \\ 0 & -1 & 0 & 0 \\ -1 & 0 & 0 & 0 \\ 0 & 0 & 0 & 0 \\ 0 & 0 & 0 & 0 \\ 0 & -1 & 0 & 0 \end{pmatrix} \text{ and } Q_1^- = \begin{pmatrix} 0 & 1 & 0 & 0 \\ 0 & 0 & 1 & 0 \\ 0 & 0 & 1 & 0 \\ 0 & 0 & 0 & 0 \\ 0 & 0 & 0 & 0 \\ 0 & 0 & 0 & 1 \end{pmatrix}$$

We now deduce that $A_{11} = \begin{pmatrix} 1 & -1 & 0 & 0 \\ 0 & 1 & -1 & 0 \\ 1 & 0 & -1 & 0 \\ -1 & -5 & 6 & 0 \\ 0 & 0 & 1 & -1 \\ 0 & 1 & 0 & -1 \end{pmatrix}$, $A_{12} = \begin{pmatrix} -1 \\ -1 \\ -1 \\ 4 \\ -1 \\ -1 \end{pmatrix}$, $A_{21} = \begin{pmatrix} -1 & 1 & 0 & 0 \\ 0 & -1 & 1 & 0 \\ -1 & 0 & 1 & 0 \\ 0 & 0 & 0 & 0 \\ 0 & 0 & 0 & 0 \\ 0 & -1 & 0 & 1 \end{pmatrix}$ and $A_{22} = \begin{pmatrix} 1 \\ 1 \\ 1 \\ 0 \\ 0 \\ 1 \end{pmatrix}$ □

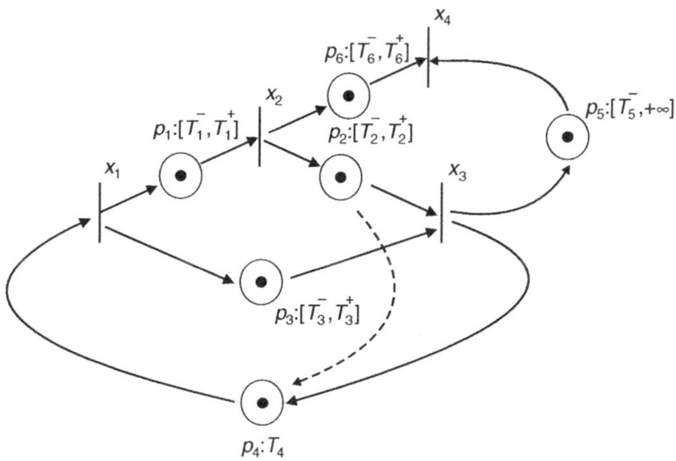

Figure 3.1. *P-time event graph with interdependent residence durations (dotted line)*

3.2.6. *Existence of a 1-periodic behavior*

After obtaining a linear inequality system having the form $A.x \leq b$, we prove the existence of a 1-periodic behavior using a well-known theorem in linear programming. As the vector b is finite, we assume that each infinite bound $T_l^+ = +\infty$ is replaced by an arbitrarily finite large size that neutralizes the constraint $x_i(k) - x_j(k - m_l) \leq T_l^+ = +\infty$.

COROLLARY 3.1.– Farkas' lemma (variant), Corollary 7.1.e in [SCH 87]. Let A be a matrix and b a vector. Then, the system $A \times x \leq b$ of linear inequalities has a solution x, if and only if $\gamma \times b \geq 0$ for each row vector $\gamma \geq 0$ with $\gamma \times A = 0$.

THEOREM 3.1.– *System [3.1] with $m = 1$ can follow a 1-periodic behavior for a given cycle time λ, if and only if, for each row vector $\lambda \geq 0$ with*

$$\gamma \times \begin{pmatrix} A_{11} \\ A_{21} \end{pmatrix} = 0 \qquad [3.6]$$

$$\gamma \times \begin{pmatrix} -T^- \\ T^+ \end{pmatrix} \geq 0 \text{ } \textit{if} \text{ } \gamma \times \begin{pmatrix} A_{12} \\ A_{22} \end{pmatrix} = 0 \qquad [3.7]$$

and the following lower and upper bounds are consistent: γ is associated with the lower bound of λ

$$\frac{\gamma \times \begin{pmatrix} -T^- \\ T^+ \end{pmatrix}}{\gamma \times \begin{pmatrix} A_{12} \\ A_{22} \end{pmatrix}} \text{ } \textit{if} \text{ } \gamma \times \begin{pmatrix} A_{12} \\ A_{22} \end{pmatrix} < 0 \qquad [3.8]$$

γ is associated with the upper bound of λ

$$\frac{\gamma \times \begin{pmatrix} -T^- \\ T^+ \end{pmatrix}}{\gamma \times \begin{pmatrix} A_{12} \\ A_{22} \end{pmatrix}} \text{ } \textit{if} \text{ } \gamma \times \begin{pmatrix} A_{12} \\ A_{22} \end{pmatrix} > 0 \qquad [3.9]$$

PROOF.– From Farkas' lemma, we can deduce that the system

$$\begin{pmatrix} A_{11} & A_{12} \\ A_{21} & A_{22} \end{pmatrix} \cdot \begin{pmatrix} x \\ \lambda \end{pmatrix} \leq \begin{pmatrix} -T^- \\ T^+ \end{pmatrix}$$

$\begin{pmatrix} A_{11} \\ A_{21} \end{pmatrix}.x \leq \begin{pmatrix} -T^- \\ T^+ \end{pmatrix} - \begin{pmatrix} A_{12} \\ A_{22} \end{pmatrix}.\lambda$ has a solution x, if and only if $\gamma \times (\begin{pmatrix} -T^- \\ T^+ \end{pmatrix} - \begin{pmatrix} A_{12} \\ A_{22} \end{pmatrix}.\lambda) \geq 0$ for each row vector $\gamma \geq 0$ with $\gamma \times \begin{pmatrix} A_{11} \\ A_{21} \end{pmatrix} = 0$.

So, $\gamma \times \begin{pmatrix} -T^- \\ T^+ \end{pmatrix} \geq \gamma \times \begin{pmatrix} A_{12} \\ A_{22} \end{pmatrix}.\lambda = \lambda \times \gamma \times \begin{pmatrix} A_{12} \\ A_{22} \end{pmatrix}$. The determination of λ leads to the upper bound $\dfrac{\gamma \times \begin{pmatrix} -T^- \\ T^+ \end{pmatrix}}{\gamma \times \begin{pmatrix} A_{12} \\ A_{22} \end{pmatrix}} \geq \lambda$ if $\gamma \times \begin{pmatrix} A_{12} \\ A_{22} \end{pmatrix} > 0$, the lower bound $\dfrac{\gamma \times \begin{pmatrix} -T^- \\ T^+ \end{pmatrix}}{\gamma \times \begin{pmatrix} A_{12} \\ A_{22} \end{pmatrix}} \leq \lambda$ if $\gamma \times \begin{pmatrix} A_{12} \\ A_{22} \end{pmatrix} < 0$, and $\gamma \times \begin{pmatrix} -T^- \\ T^+ \end{pmatrix} \geq 0$ if $\gamma \times \begin{pmatrix} A_{12} \\ A_{22} \end{pmatrix} = 0$.

For any vectors γ_1, γ_2 satisfying $\gamma \times \begin{pmatrix} A_{12} \\ A_{22} \end{pmatrix} = 0$ and corresponding to the lower and upper bounds, respectively, we have $\dfrac{\gamma_1 \times \begin{pmatrix} -T^- \\ T^+ \end{pmatrix}}{\gamma_1 \times \begin{pmatrix} A_{12} \\ A_{22} \end{pmatrix}} \leq \dfrac{\gamma_2 \times \begin{pmatrix} -T^- \\ T^+ \end{pmatrix}}{\gamma_2 \times \begin{pmatrix} A_{12} \\ A_{22} \end{pmatrix}}$ as λ is defined by these bounds. □

Therefore, the theorem brings a way of checking the consistency of [3.4] and the existence of a 1-periodic trajectory for the extended P-time event graphs. Moreover, this result gives a frame of the cycle time given by the maximum of the lower bounds and the minimum of the upper bounds. Note that the existence of a solution depends on λ in the two last relations contrary to the first one.

REMARK.– Note that the vector y is not a P-invariant as its dimension is 2. $\mid P \mid$. It is a generalized P-invariant that depends on the incidence matrix expressed in A_{11} and A_{21} and on the affine functions of the interdependent residence durations.

Graphical interpretation. In the case of a P-time event graph, equation [3.6] becomes $\gamma \times \begin{pmatrix} W \\ -W \end{pmatrix} = 0$ and a graph can be associated with matrix $\begin{pmatrix} W \\ -W \end{pmatrix}$: as W is the incidence matrix of an event graph where each row contains the two entries 1 and -1 (corresponding to columns i and j), each row l of this matrix can be associated with an arc coming from vertex j to vertex i. Following relation [3.5], we can associate the time value $-T_l^-$ (and the marking value $-(M_0)_l = (A_{12})_l$) with the arc l coming from vertex i to vertex j for the upper matrix W and the time value T_l^+ (and the marking value $(M_0)_l = (A_{22})_l$) with the relevant arc coming from vertex j to vertex i for the lower matrix $-W$. As the non-null entries of the vector $\gamma \geq 0$ can be normalized to 1, the vector γ highlights the rows and their bounds that correspond to the minimal and maximal solutions. It is well known that each vector γ defines a circuit in the associated graph and that the set of these circuits describes the minimal and maximal critical structures that determine the bounds of the cycle time.

However, this graphical interpretation does not hold for a P-time event graph with interdependent residence durations: we have $A_{11} \neq W$ and $A_{21} \neq -W$ and a row cannot be associated with an arc as it can contain more than two entries. We can define a different associated graph expressing the connections between the rows and the columns of matrix $\begin{pmatrix} A_{11} \\ A_{21} \end{pmatrix}$ as follows: each row is associated with a specific element (a black dot, for instance) linked to the different vertices corresponding to the columns (a vertical line like a transition). Moreover, the entries of vector γ cannot be normalized to 1 in general, which implies that the selection of the rows by the non-null entries of vector γ, and the relevant selected subgraph, presents a ponderation. If we neglect this ponderation, the selected rows do not define a circuit but a more complex structure. In fact, each vector γ expresses a dependence between the rows of the relevant submatrix. Following relation [3.4], we can associate the pair $(-T_l^-, (A_{12})_l)$ with row l of the upper matrix A_{11} and the pair $(T_l^+, (A_{22})_l)$ with row l of the lower matrix A_{21}. The following example shows that only a subset of the vectors γ corresponds to a circuit, in the Petri net or in the associated graph.

3.2.7. *Example continued*

More general than the matrices obtained in the studies [DEC 07b, DEC 08], matrices Q_1^- and Q_0^- are almost but are not the incidence matrices W^+ and W^- of an event graph as the fourth row contains two non-null entries (see section 3.2.5).

We also have $A_{11} \neq W$ and $A_{21} \neq -W$. The analysis of the independent row vectors γ

$$\begin{pmatrix} 0.316 & 2.025 & 0.025 & 0.341 & 0 & 0 & 0 & 0 & 0 & 0 & 0 & 0 \\ 0 & 1 & 0 & 0 & 1 & 0 & 0 & 0 & 0 & 0 & 0 & 1 \\ 0 & 2.5 & 0.5 & 0.5 & 0 & 0 & 0 & 0 & 0 & 0 & 0 & 0 \\ 0 & 0 & 2 & 0.33 & 0 & 0 & 1.66 & 0 & 0 & 0 & 0 & 0 \end{pmatrix}$$

highlights the following relevant structures:

γ	Structure	Bound
γ_1	$-T_1^-, -T_2^-, -T_3^-$ and b	Lower bound: 8.075
γ_2	$-T_2^-, -T_5^-$ and T_6^+	Lower bound: 9.5
γ_3	$-T_2^-, -T_3^-$ and b	Lower bound: 9.5
γ_4	$-T_3^-, b$ and T_1^+	Upper bound: 13.66

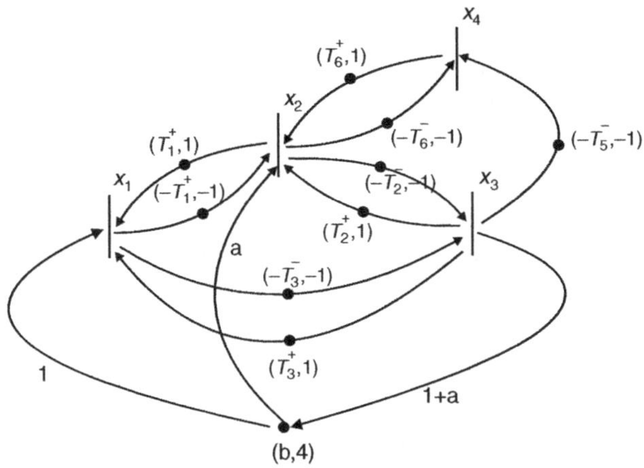

Figure 3.2. *Associated graph: each row corresponds to a black dot while each column is expressed by a vertical line*

The structure selected by γ_2 ($-T_2^-$, $-T_5^-$ and T_6^+) corresponds to a circuit in the associated graph contrary to the structures selected by γ_1, γ_3 and γ_4, which present non-disjoint circuits. Note that the structure selected by γ_1 is not irreducible as the relevant submatrix has four rows and three columns. Therefore, the usual theorem, which gives the cycle time based on the circuits, does not hold. We obtain $\max(8.075, 9.5, 9.5) \leq \lambda \leq \min(13.66)$ and the interval of the possible values is [9.5, 13.66].

Note also that the vectors γ_2 and γ_4, which define a lower bound and an upper bound, respectively, of the cycle time, select a combination of lower and upper bounds of temporizations and not only a set of lower (respectively, upper) bounds of temporizations. □

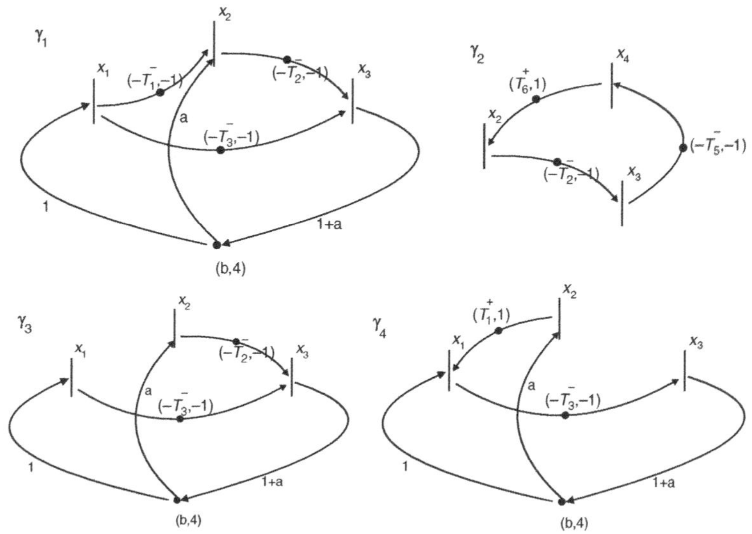

Figure 3.3. *Substructures*

3.3. Optimization

3.3.1. *Approach 1*

Let us define the criterion of our problem. A classical objective can be the determination of the minimal and maximal cycle times like in many studies. The objective can also be the maximization of the residence duration of some tasks like the use of machines. It can be the minimization of the residence duration of some other tasks like the working time of the employees. The sets of the places are denoted by P_{machines} and $P_{\text{employees}}$, respectively. Therefore, a more general criterion is $Cr.\begin{pmatrix} x \\ \lambda \end{pmatrix} = \alpha_1 \cdot \sum\limits_{p_l \in P_{\text{machines}}} (x_{p_l^\bullet}(k) - x_{^\bullet p_l}(k - m_l)) + \alpha_2 \cdot \sum\limits_{p_l \in P_{\text{employees}}} (x_{p_l^\bullet}(k) - x_{^\bullet p_l}(k - m_l)) + \alpha_3 \cdot \lambda$ where α_1, α_2 and α_3

are ponderation coefficients. We can focus on the optimization of the residence durations by taking $\alpha_1 > 0$, $\alpha_2 < 0$ and $\alpha_3 = 0$: the goal is the maximization of the use of the machines and the minimization of the working time of the employees during the pattern λ. The periodic behavior leads to the repetition of the obtained gain.

As $x_{p_l^\bullet}(k) - x_{\bullet p_l}(k - m_l) = x_{p_l^\bullet}(k) - x_{\bullet p_l}(k) + m_l \cdot \lambda$, a more simple way of writing this is as follows:

$Cr \cdot \begin{pmatrix} x \\ \lambda \end{pmatrix} = \alpha_1 \cdot \sum_{p_l \in P_{\text{machines}}} (x_{p_l^\bullet}(k) - x_{\bullet p_l}(k)) + \alpha_2 \cdot \sum_{p_l \in P_{\text{employees}}} (x_{p_l^\bullet}(k) - x_{\bullet p_l}(k)) + \alpha_3' \cdot \lambda$ where $\alpha_3' = \alpha_3 + \alpha_1 \cdot \sum_{p_l \in P_{\text{machines}}} m_l + \alpha_2 \cdot \sum_{p_l \in P_{\text{employees}}} m_l$.

A finite optimal solution yields a realistic trajectory and the smallest or largest vector x is not bounded up to now. The solution is to limit the minimization (respectively, maximization) by a minorant (respectively, a majorant) taken finished. The minorant of state x (respectively, majorant) is noted Lb (respectively, noted Ub). The non-decrease of the trajectories can be obtained by adding the relation $x(k+1) \geq x(k)$ in the model and a consequence is $\lambda \geq 0$.

Under the condition of compatibility of the system of constraints, the optimal solution is given by the following linear programming problems.

$\min Cr \cdot \begin{pmatrix} x \\ \lambda \end{pmatrix}$ under constraint:

$$\begin{pmatrix} -I & 0 \\ A_{11} & A_{12} \\ A_{21} & A_{22} \end{pmatrix} \cdot \begin{pmatrix} x \\ \lambda \end{pmatrix} \leq \begin{pmatrix} -Lb \\ -T^- \\ T^+ \end{pmatrix} \qquad [3.10]$$

$\max Cr \cdot \begin{pmatrix} x \\ \lambda \end{pmatrix}$ under constraint:

$$\begin{pmatrix} I & 0 \\ A_{11} & A_{12} \\ A_{21} & A_{22} \end{pmatrix} \cdot \begin{pmatrix} x \\ \lambda \end{pmatrix} \leq \begin{pmatrix} Ub \\ -T^- \\ T^+ \end{pmatrix} \qquad [3.11]$$

REMARK.– The rows corresponding to an arbitrary large upper temporization T_i^+ can be removed as they do not modify the calculation.

The reduction to the optimization of the cycle time of a P-time event graph without interdependent residence durations is as follows. The minimization/maximization of only the cycle time λ is obtained with $\alpha_3 > 0$, $\alpha_1 = 0$ and $\alpha_2 = 0$. The constraints of a P-time event graph without interdependent residence durations are given by $A_{11} = W = -A_{21}$ and $A_{12} = -M_0 = -A_{22}$. Note that each line of the constraint contains three variables per inequality (λ and two variables x_i), which is not the general case.

3.3.2. *Example continued*

Below we take $Lb = (0 \quad 0 \quad 0 \quad 0)^t$ and $Ub = (20 \quad 20 \quad 20 \quad 20)^t$.

Criterion 1. We focus on the optimization of the cycle time λ: $Cr = (0 \quad 0 \quad 0 \quad 0 \quad 1)$.

For $\lambda_{\min}$, we obtain $(x_1 \quad x_2 \quad x_3 \quad x_4 \quad \lambda)^t = (8.5 \quad 6.5 \quad 0 \quad 2 \quad 9.5)^t$ with the criterion value $f = 9.5$. The minimal cycle time is $\lambda_{\min} = 9.5$. So, the relevant trajectory is $x(0) = (8.5 \quad 6.5 \quad 0 \quad 2)^t \rightarrow x(1) = (18 \quad 16 \quad 9.5 \quad 11.5)^t \rightarrow x(2) = (27.5 \quad 25.5 \quad 19 \quad 21)^t \rightarrow \ldots$

For $\lambda_{\max}$, we obtain $(x_1 \quad x_2 \quad x_3 \quad x_4 \quad \lambda)^t = (20 \quad 16.33 \quad 7.33 \quad 7.66 \quad 13.66)^t$ with $f = 13.66$. The maximal cycle time is $\lambda_{\max} = 13.66$.

Criterion 2. Now we want: (1) to minimize the cycle time λ; (2) to minimize $x_3(k) - x_2(k-1)$, which expresses the cooking of a product; (3) to maximize $x_2(k) - x_1(k-1)$. So, we take $Cr.\begin{pmatrix} x \\ \lambda \end{pmatrix} = \alpha_1.(x_{p_2^\bullet}(k) - x_{\bullet p_2}(k - m_2)) + \alpha_2. \sum_{p_l \in P_{\text{employees}}} (x_{p_1^\bullet}(k) - x_{\bullet p1}(k - m_1)) + \alpha_3.\lambda$ with the ponderations $\alpha_1 = 2$, $\alpha_2 = -2$ and $\alpha_3 = 1$. We obtain $Cr. = (-\alpha_2 \quad \alpha_2 - \alpha_1 \quad \alpha_1 \quad 0 \quad \alpha_1 + \alpha_2 + 1) = (2 \quad -4 \quad 2 \quad 0 \quad 1)$.

The minimization of the above criterion gives $(x_1 \quad x_2 \quad x_3 \quad x_4 \quad \lambda)^t = (9 \quad 8 \quad 0 \quad 0.5 \quad 11)^t$ with $f = -3$. The cycle time $\lambda = 11$ is greater than the minimal value $\lambda_{\min} = 9.5$. The analysis of the terms $x_3(k) - x_2(k-1)$ and $x_2(k) - x_1(k-1)$ for the two criteria shows that the term $x_2(k) - x_1(k-1)$ has been increased (7.5 for criterion 1 and 10 for criterion 2) while the other one is identical.

The maximization of criterion 2 leads to the same results as criterion 1: $(x_1 \quad x_2 \quad x_3 \quad x_4 \quad \lambda)^t = (20 \quad 16.33 \quad 7.33 \quad 7.66 \quad 13.66)^t$ but with $f = 3$. □

3.3.3. *Approach 2*

In approach 1, the variables used are the cycle time and the first date of shooting of each transition. We now apply the principle of duality allowing the replacement of these variables by a new variable denoted by γ. This variable γ gives information on the part of the system associated with the optimal cycle time (see also sections 3.2.6 and 3.2.7). Moreover, this principle generates a new algorithm.

3.3.3.1. *The two dual forms*

Let us first recall the dual problems.

$$\text{Problem (P)} \min_{y \in \mathbb{R}^n} y.b \text{ under } y.A = c \text{ and } y \geq 0$$

and

Problem (D) $\max x \in \mathbb{R}^m$ $c.x$ under $A.x \leq b$

Let us consider the two problems of approach 1. This corresponds to the form (D) with the following correspondences for the minimization [3.10]:

$$A = \begin{pmatrix} -I & 0 \\ A_{11} & A_{12} \\ A_{21} & A_{22} \end{pmatrix}, b = \begin{pmatrix} -Lb \\ -T^- \\ T^+ \end{pmatrix}, x = \begin{pmatrix} x \\ \lambda \end{pmatrix} \text{ and } c = -Cr$$

For the maximization [3.11], we have

$$A = \begin{pmatrix} I & 0 \\ A_{11} & A_{12} \\ A_{21} & A_{22} \end{pmatrix}, b = \begin{pmatrix} Ub \\ -T^- \\ T^+ \end{pmatrix}, x = \begin{pmatrix} x \\ \lambda \end{pmatrix} \text{ and } c = +Cr$$

We thus obtain the following dual problem that is also a linear problem of programming:

$$\min \ \gamma.\left(-Lb^t \quad -(T^-)^t \quad (T^+)^t\right)^t \text{ under } \gamma.\begin{pmatrix} -I & 0 \\ A_{11} & A_{12} \\ A_{21} & A_{22} \end{pmatrix} =$$

$-Cr$ with $\gamma \geq 0$ and

$$\min \ \gamma.\left(Ub^t \quad -(T^-)^t \quad (T^+)^t\right)^t \text{ under } \gamma.\begin{pmatrix} I & 0 \\ A_{11} & A_{12} \\ A_{21} & A_{22} \end{pmatrix} = Cr$$

with $\gamma \geq 0$.

The following theorem will be useful. The set $X^{ad} = \{x \in \mathbb{R}^m | A.x \leq b\}$ is the set of admissible solutions to the dual problem (D). In the same way, $Y^{ad} = \{y \in (\mathbb{R}^+)^n | y.A = c\}$ for problem (P).

THEOREM 3.2 (Chapter 4 in [CUL 94]).–

1) if $y \in Y^{ad}$ *and* $x \in X^{ad}$*, then* $y.b \geq c.x$

2) if $\bar{y} \in Y^{ad}$, $\bar{x} \in X^{ad}$ and $y.b = c.x$, then $\bar{y}$ and $\bar{x}$ are the optimal solutions to (P) and (D), respectively.

We now focus on the optimization of the cycle time in the case of an extended P-time event graph. The particular case of the P-time event graphs will be considered in the remarks.

3.3.3.2. *Optimization of the cycle time*

The application of the theorem of duality will show that the optimal solution of approach 2 (problem P) also gives the optimal cycle time. Let $b_1^t = (-Lb^t \quad (-T^-)^t \quad (T^+)^t)$ and $b_2^t = (Ub^t \quad (-T^-)^t \quad (T^+)^t)$ be the two vectors b in problems [3.10] and [3.11]. Let $\begin{pmatrix} x \\ \lambda \end{pmatrix} \in X^{ad}$ and $y \in Y^{ad}$, $\begin{pmatrix} x_{\min} \\ \lambda_{\min} \end{pmatrix}$ and $y_{\min}$ ($\begin{pmatrix} x_{\max} \\ \lambda_{\max} \end{pmatrix}$ and $y_{\max}$, respectively) be the optimal solutions to the problem of minimization (maximization, respectively).

So, we have the following result when the optimization of only the cycle time is considered.

THEOREM 3.3.– *Let* $Cr = (0 \ldots 0 \quad 1)$

$$-\gamma.b_1 \leq \lambda \text{ and } \lambda_{\min} = -\gamma_{\min}.b_1$$

$$+\lambda \leq \gamma.b_2 \text{ and } \lambda_{\max} = +\gamma_{\max}.b_2$$

PROOF.– The first point of the duality theorem indicates that: $\gamma.b \geq c.x$ where, for problem [3.10], $b_1^t.\gamma^t = (-Lb^t \quad -(T^-)^t \quad (T^+)^t).\gamma^t \geq (0 \ldots 0 \quad -1)\begin{pmatrix} x \\ \lambda \end{pmatrix} = -\lambda$.

Moreover, the associated optimal values are equal (point 2) if the two problems admit a finite optimal solution. Similarly, we have

$$b_2^t.\gamma^t = (Ub^t \quad (-T^-)^t \quad (T^+)^t).\gamma^t \geq (0 \ldots 0 \quad +1)\begin{pmatrix} x \\ \lambda \end{pmatrix} = +\lambda$$

□

3.3.3.3. *Improvement 1*

The study of section 3.2.6 shows that the bounds of the cycle time depend only on the matrices, which define the P-time event graph, and not on the vectors Lb and Ub. As a consequence, we can simplify approaches 1 and 2 if we consider only the optimization of the cycle time ($Cr = (0\ldots0 \quad -1)$), by taking $Lb = 0$ (respectively, $Ub = 0$). This hypothesis leads to a shift on the first dates of shooting of the transitions in approach 1.

We denote $\gamma = (\gamma_1 \quad \gamma_2)$ where γ_1 has the dimension of x.

PROPERTY 3.2.– The problem (P) can be rewritten in the following form:

$$min\left(-(T^-)^t \quad (T^+)^t\right).\gamma_2^t$$

under:

$$\begin{cases} \gamma_2.\begin{pmatrix} A_{11} \\ A_{21} \end{pmatrix} \geq 0 \\ \gamma_2.\begin{pmatrix} A_{12} \\ A_{22} \end{pmatrix} = -1 \\ \gamma_2 \geq 0 \end{cases} \qquad [3.12]$$

PROOF.– As $Lb = 0$, only the subvector γ_2 is optimized and we obtain $\min \gamma.(0 \quad -(T^-)^t \quad (T^+)^t)^t$ under $\gamma.\begin{pmatrix} -I & 0 \\ A_{11} & A_{12} \\ A_{21} & A_{22} \end{pmatrix} = -Cr$ with $\gamma \geq 0$. Recall that $Cr = (0\ldots0 \quad 1)$.

The problem becomes:

$$\gamma_1.(-I \quad 0) + \gamma_2.\begin{pmatrix} A_{11} & A_{12} \\ A_{21} & A_{22} \end{pmatrix} = -Cr = (0\ldots0 \quad -1)$$

So, we can deduce that $\gamma_2.\begin{pmatrix}A_{12}\\A_{22}\end{pmatrix} = -1$ and $\gamma_2.\begin{pmatrix}A_{11}\\A_{21}\end{pmatrix} \geq (0\ldots0)$ as $\gamma_1 \geq 0$

The vector γ_1, which is not optimized, is obtained with $\gamma_1. = \gamma_2.\begin{pmatrix}A_{11} & A_{12}\\A_{21} & A_{22}\end{pmatrix} + Cr.$ □

REMARK.– In the case of a P-time event graph, we can easily deduce the matrices by using the relations $A_{11} = W = -A_{21}$ and $A_{12} = -M_0 = -A_{22}$.

3.3.3.4. *Improvement 2*

The following theorem will analyze the first relationship [3.12].

THEOREM 3.4 (Stiemke 1915, [SCH 87]).– *For a matrix A, the following cases are mutually exclusive from each other.*

– *Case 1.* $Ax = 0$, $x > 0$ *has a solution* x.

– *Case 2.* $y.A \geq 0$ *and* $y.A \neq 0$ *has a solution* y.

PROPERTY 3.3.– In relation [3.12] of property 3.2, $\gamma_2.\begin{pmatrix}A_{11}\\A_{21}\end{pmatrix} \geq 0$ becomes $\gamma_2.\begin{pmatrix}A_{11}\\A_{21}\end{pmatrix} = 0$ if $\begin{pmatrix}A_{11}\\A_{21}\end{pmatrix}.x = 0$, $x > 0$ has a solution x.

PROOF.– The application of the Stiemke's theorem entails that case 2 has no solution: the system $\gamma.A \geq 0$ and $\gamma.A \neq 0$ has no solution γ. On the other hand, the condition $\gamma.A \geq 0$ in [3.12] must be checked in order to calculate the cycle time. The only possibility for the system is $\gamma.A = 0$. □

REMARK.– In an event graph, there is $x > 0$ as $W.x = 0$ as each line of W has exactly one coefficient 1 and -1 (case 1). It can be concluded that the events graph is "consistent", that is there is a sequence of shooting and an initial marking that allows us to find the same marking under the condition that each transition is fired at least once (see page 567 of [MUR 89]). For a P-time event graph, we have case 1 as $A_{11} = W = -A_{21}$.

3.3.3.5. *Example continued*

Approach 2

Stiemke's theorem that can be applied as $x = (1 \quad 1 \quad 1 \quad 1)^t$ is a solution to $\begin{pmatrix} A_{11} \\ A_{21} \end{pmatrix} .x = 0,\, x > 0$. We can use the simplified dual approach 2 with relation [3.12] with $\gamma_2 . \begin{pmatrix} A_{11} \\ A_{21} \end{pmatrix} = 0$.

Dual $\lambda_{\min}$

$\gamma = (0 \quad 2.5 \quad 0.5 \quad 0.5 \quad 0 \quad 0 \quad 0 \quad 0 \quad 0 \quad 0 \quad 0 \quad 0)$ with $f = 9.5$.

Dual $\lambda_{\max}$

$\gamma = (0 \quad 0 \quad 2 \quad 0.33 \quad 0 \quad 0 \quad 1.66 \quad 0 \quad 0 \quad 0 \quad 0 \quad 0)$ with $f = 13.66$. These vectors, respectively, correspond to γ_3 and γ_4 calculated in section 3.2.6. The same interval of the possible values for the cycle time is obtained ($\lambda \in [9.5, 13.66]$). □

3.4. Conclusion

In this chapter, we generalized the model of the P-time event graph [KHA 97] by allowing the dependence between the time durations and considering its performance analysis. We first showed that this new model can be completely described by a polyhedron $Ax \leq b$. Different to the matrices of a P-time graph where there are correspondences between matrix A and incidence matrices, the matrices can have some entries that depend on time parameters and each row can contain more than two entries. Therefore, the system cannot be completely analyzed with a classical graph theory like in the case of a timed event graph or a P-time event graph: we propose the use of linear programming in this chapter.

An objective of this chapter is the analysis of the performance with a criterion function of the cycle time and of the time durations associated with places: we rewrite the objective in the form of a linear programming problem whose resolution generates the optimal initial dates and cycle time of the problem. The direct application of these initial dates leads to a 1-periodic behavior of the system with the calculated cycle time.

Using the principle of duality, we also proposed a second approach that can calculate the optimal cycle times. The main advantage is that the different results and conditions only depend on the constants of the system and not on the dates of firing. We show in the example that the cycle time can be associated with structures that are more general than circuits in the associated graph.

The application of Farkas' lemma gives conditions of consistency of the problem and an interval limiting the cycle time: this result shows that the concept of cycle time holds on the new model as the conditions depend on the characteristics of the system only. Finally, the application of Stiemke's theorem allows a simplification under an algebraic condition. This last one is satisfied if the study is limited to P-time event graphs and to the illustrative example that contains a dependence between different time durations.

3.5. Appendix

Now let us express system inequalities [3.1] on a reduced horizon. Such a form will simplify the calculations. The objective is to establish an equivalent model such that each place of the graph contains zero or only one token initially.

As a place contains a maximum number of m tokens, the general idea is to split each place containing m tokens into m places, where each place contains only one token. A systematic procedure is as follows.

Let us introduce a new variable X, that is:

$X(k) = \begin{pmatrix} X_0(k) & .... & X_i(k) & .... & X_{m-1}(k) \end{pmatrix}$ with $X_i(k) = x(k-m+i+1)$. By construction, we have: $X_{m-1}(k) = x(k)$ and $X_i(k) = X_{i+1}(k-1)$ for i going from 0 to $m-2$.

So, system [3.1] becomes:

$$\begin{pmatrix} Q_1' & Q_0' \end{pmatrix} \times \begin{pmatrix} X(k-1) \\ X(k) \end{pmatrix} \leq \begin{pmatrix} -T^- \\ T^+ \end{pmatrix}$$

where $Q_1' = \begin{pmatrix} Q_m & 0 & \ldots & \ldots & 0 \end{pmatrix}$

and $Q_0' = \begin{pmatrix} Q_{m-1} & Q_{m-2} & \ldots & Q_1 & Q_0 \end{pmatrix}$.

We complete the system with the following relations:

$$\begin{cases} +X_{i+1}(k-1) - X_i(k) \leq 0 \\ -X_{i+1}(k-1) + X_i(k) \leq 0 \end{cases}$$

for $i = 0$ to $m-2$ and we obtain

$$\begin{pmatrix} O_{01}^- & O_{00}^- \\ O_{01}^+ & O_{00}^+ \end{pmatrix} \times \begin{pmatrix} X(k-1) \\ X(k) \end{pmatrix} \leq \begin{pmatrix} 0 \\ 0 \end{pmatrix}$$

where matrix $O_{01}^- = -O_{01}^+$ of dimension $((m-1) \times |TR| \times m)$ is a subdiagonal of identity matrices immediately above the main diagonal, while the matrix $O_{00}^- = -O_{00}^+$ is a diagonal of negative identity matrices.

Finally, the concatenation of the two systems gives the algebraic form:

$$\begin{pmatrix} H^- \\ H^+ \end{pmatrix} \cdot \begin{pmatrix} X(k-1) \\ X(k) \end{pmatrix} \leq \begin{pmatrix} -T^- \\ 0 \\ T^+ \\ 0 \end{pmatrix}$$

where $H^- = \begin{pmatrix} Q_1'^- & Q_0'^- \\ O_{01}^- & O_{00}^- \end{pmatrix}$ and $H^+ = \begin{pmatrix} Q_1'^+ & Q_0'^+ \\ O_{01}^+ & O_{00}^+ \end{pmatrix}$.

4

Control with Specifications

The purpose of this chapter is the optimal control synthesis of a timed event graph when the state and control trajectories should follow the specifications defined by an interval model. The problem is reformulated in the fixed-point form and the spectral theory gives the conditions of the existence of a solution.

4.1. Introduction

In this chapter, we focus on the control problem of timed event graphs defined as follows. In a timed event graph, some events are stated as controllable, meaning that the corresponding transitions (input) may be delayed from firing until some arbitrary time provided by a supervisor. Moreover, we assume that the timed event graph must follow specifications defined by a second model. As the specifications can be incoherent or too restrictive with respect to the timed event graph, the first objective is to determine whether there exist control actions that will restrict the system to that behavior. If the trajectories of the timed event graph can follow additional specifications, the second objective is to determine the greatest input in order to obtain the desired behavior defined by the static constraints (the desired output) and the dynamic constraints (expressed by a second model).

In this chapter, this second model is defined by a set of time interval systems for which the time evolution is not strictly deterministic. It is described by intervals that use the operations of maximization, minimization and addition to define the lower and upper bound constraints on the dates of firing of all the transitions. P-time event graphs and time stream event graphs can be modeled algebraically by interval systems.

In this chapter (and in Chapter 5), we consider that each transition is observable: the event date of each transition firing is assumed to be available. No hypothesis is taken on the structure of the event graphs that does not need to be strongly connected. The initial marking should only satisfy the classical liveness condition and the usual hypothesis that places should be First In First Out (FIFO) is taken.

The chapter is structured as follows: the different models and the interval (min, max, +) systems are described in the following section. On the basis of a fixed-point approach, we successively present the optimal control problem and analyze the consistency of the problem, that is, the existence of a finite solution. Using the spectral theory, we analyze a more general model than the P-time event graph considered in Chapter 2. We then present an algorithm that determines the greatest control. Finally, the approach is applied to a general example. The chapter also presents an everyday example.

Let us remember that the notations are given in Chapter 2 on consistency: maximization, minimization and addition operations are denoted, respectively, by $\oplus$, $\wedge$ and $\otimes$; $\varepsilon = -\infty$ and $T = +\infty$.

4.2. Time interval systems

Discrete event dynamic systems involving synchronization can be modeled by several types of Petri net. The class of time event graphs includes P-time event graphs and time stream event graphs, both of which extend the application field of timed event graphs. For example, time stream event graphs allow the specification of synchronization requirements of multimedia applications [DIA 97] and the description of complex synchronizations. The time stream Petri nets present different types of semantic rules such as “And”, “Weak-And”, “Strong-Or”, “Or”, “Master” and their variations [DIA 01], which correspond to different temporal evolutions. We show in the following section that timed event graphs, P-time event graphs and time stream event graphs can be modeled by a new class of systems called interval systems, for which the time evolution is not strictly deterministic, but instead belongs to intervals whose lower and upper bound constraints are defined by the operations of maximization, minimization and addition.

4.2.1. *(min, max, +) algebraic models*

DEFINITION 4.1.– *[COC 99] A* (min, max, +) *function of type* $(n, 1)$ *is any function* $f\colon \mathbb{R}^n \longrightarrow \mathbb{R}^1$, *which can be written as a term in the following construction:* $f = x_1, x_2, \ldots, x_n \mid f \otimes a \mid f \wedge f \mid f \oplus f$, *where* a *is an arbitrary real number* ($a \in \mathbb{R}$).

The vertical bars separate the different ways in which terms can be recursively constructed. A (min, max, +) function of type (n, m) is any function $f\colon \mathbb{R}^n \longrightarrow \mathbb{R}^m$ such that each component f_i is a (min, max, +) function of type $(n, 1)$. The set of functions (min, max, +) of the type (n, m) is denoted by $F(n, m)$.

The evolution of the system is described by the following model, called an “interval model” or an “interval descriptor system”, where f^- and f^+ are (min, max, +) functions.

$$f^-(x(k-1), x(k), u(k)) \leq x(k) \leq f^+(x(k-1), x(k), u(k)) \quad [4.1]$$

Vectors x and u are the state and the input, respectively. Variable $x_i(k)$ (respectively, $u_i(k)$) is the date of the kth firing of transition x_i (respectively, u_i) for $k \in \mathbb{Z}$. In the same way and without loss of generality, we can also introduce the output y by

$$y(k) = C \otimes x(k) \text{ with } C_{ij} \in \mathbb{R}_{\max}$$

As the type of the system is defined by the types of function f^- and f^+, we can characterize the model by the following pair (type of f^-, type of f^+) that defines different types of system. The type ((min, max, +), (min, max, +)) naturally represents the more general mathematical case.

In the following sections, we will express the interval of firing of each transition for different time event graphs. Let us note that Event Graphs constitute a subclass of Petri nets in which each place has exactly one upstream and one downstream transition. Alleviating the presentation, the following assumption is made: every internal place contains at most one token; every input (respectively, every output) place contains no token. This form can simply be obtained by elementary transformations, such as duplication of places, or by algebraic transformations.

4.2.2. *Timed event graphs*

It is well known that the model of a timed event graph in the (max, +) dioid (see [BAC 92] for details) is as follows:

$$x(k) \geq A_1^- \otimes x(k-1) \oplus A_0^- \otimes x(k) \oplus B \otimes u(k) \qquad [4.2]$$

with $(A)_{ij}, (B)_{ij}, \in \mathbb{R}_{\max}$

This model represents non-decreasing trajectories if matrix A is simply modified such that $x(k) \geq x(k-1)$. This remark holds for every model considered in this chapter.

An equivalent model can also be obtained if the lower bound constraint defined by f^- is max-only ($f^-(x(k-1), x(k), u(k)) = A_1^- \otimes x(k-1) \oplus A_0^- \otimes x(k) \oplus B \otimes u(k)$) and the upper bound constraint $f^+()$ is infinite.

$$\begin{cases} x(k) \geq f^-(x(k-1), x(k), u(k)) = A_1^- \otimes x(k-1) \\ \qquad \oplus A_0^- \otimes x(k) \oplus B \otimes u(k) \\ x(k) \leq f^+(x(k-1), x(k), u(k)) = +\infty \end{cases} \quad [4.3]$$

Consequently, a timed event graph can also be described by an interval model.

The usual assumption that there is no extra delay for firing transitions whenever all tokens are available can be made and it can lead to the following equality.

$$x(k) = A_1^- \otimes x(k-1) \oplus A_0^- \otimes x(k) \oplus B \otimes u(k) \quad [4.4]$$

This equation also leads to another interval system whose bounds are equal and defined by:

$$\begin{cases} x(k) \geq f^-(x(k-1), x(k), u(k)) = A_1^- \otimes x(k-1) \\ \qquad \oplus A_0^- \otimes x(k) \oplus B \otimes u(k) \\ x(k) \leq f^+(x(k-1), x(k), u(k)) = A_1^- \otimes x(k-1) \\ \qquad \oplus A_0^- \otimes x(k) \oplus B \otimes u(k) \end{cases} \quad [4.5]$$

4.2.3. *P-time event graphs*

Let ST_i be the set of input transitions of transition i. Already described in the previous chapters, the model of a P-time event graph is as follows:

$$\bigoplus_{j \in ST_i} (x_j(k-m_j) + T_j^-) \leq x_i(k) \leq \bigwedge_{j \in ST_i} (x_j(k-m_j) + T_j^+)$$

where m_j is the initial marking of place p_j. The lower bound (respectively, the upper bound) is a (max, +) function (respectively (min, +) function) and as a result, a P-time event graph is an example of an interval system whose type is ((max, +), (min, +)). If $m_j = 1$ for every internal place j, this is defined by

$$\begin{cases} x(k) \geq f^-(x(k-1), x(k), u(k)) = A_1^- \otimes x(k-1) \\ \qquad \oplus A_0^- \otimes x(k) \oplus B^- u(k) \\ x(k) \leq f^+(x(k-1), x(k), u(k)) = A_1^+ \\ \qquad \odot x(k-1) \wedge A_0^+ \odot x(k) \wedge B^+ \odot u(k) \end{cases} \quad [4.6]$$

REMARK.– If every upper bound of places equals infinite, the model of a timed event graph is obtained [4.2].

4.2.4. *Time stream event graphs*

DEFINITION 4.2 (Time stream event graph [COU 96] [DIA 97]).– *Let I_j be a set of upstream arcs of a transition j and P_j be the corresponding set of upstream places. A time stream event graph is an event graph such that:*

a) an interval $[T_i^-, T_i^+] \in (\mathbb{R}^+ \cup \{0\}) \times (\mathbb{R}^+ \cup \{+\infty\})$ is associated with each arc $a_i \in I_j$ (usually defined on $\mathbb{Q}^+$, the limits of intervals are generalized to $\mathbb{R}^+$, which does not introduce new difficulties);

b) a special semantic rule of firing associated with each transition is defined below.

Consider one out-going arc from a given place. When a token is received by that place at time x, the token should remain in the place for a period of time defined by a value within the range $[x + T^-, x + T^+]$ associated with the arc. As the firing time of a transition that has more than one arc depends on the nature of the processes which will be synchronized, different semantic rules of firing may be associated with a transition.

DEFINITION 4.3.– *For a transition i, let I_i denote a set of upstream arcs and P_i denote the corresponding set of upstream* *places. A transition i of the type "And" and "Weak-And" is fired at time x_i if and only if the two following conditions are satisfied:*

a) Transition i is enabled for the current marking: every upstream place j of P_i contains at least one token. Let x_j be the entrance date of the token that is also the date of firing of the upstream transition of this place.

b) For the semantic rule And, *the value of x_i is as follows: $(x_j + T_j^-) \leq x_i \leq (x_j + T_j^+)$ for every upstream place $p_j \in P_i$ and arc $a_j \in I_i$. The model can be written as follows: if m_j is the initial marking of the place p_j, the following expression can be written for each transition:*

$$\bigoplus_{j \in P_i} (x_j(k - m_j) + T_j^-) \leq x_i(k) \leq \bigwedge_{j \in P_i} (x_j(k - m_j) + T_j^+)$$

if the semantic rule is "And".

Let us note that the inequalities of the P-time event graph correspond to semantic rule "And" of the time stream event graph.

For the semantic rule Weak-And, *the value of x_i is as follows: $(x_i + T_i^-) \leq x_i$ for every upstream place $p_j \in P_i$ and arc $a_j \in I_i$ and $\exists j \in P_{i,}\ x_i \leq (x_j + T_j^+)$ (the firing may wait until the last time interval).*

$$\bigoplus_{j \in P_i} (x_j(k - m_j) + T_j^-) \leq x_i(k) \leq \bigoplus_{j \in P_i} (x_j(k - m_j) + T_j^+)$$

if the semantic rule is "Weak-And".

Therefore, for these two semantic rules, the upper bound is either (min, +) or (max, +). Now, we will describe the time stream event graph in the form of an interval system. The following definitions allow the introduction of a standard description.

Let $f \in F(n,n)$. A subset $S \subset F(n,n)$ is said to be a max-representation of f if S is a finite set of (max,+) functions such that $f = \bigwedge_{g \in S} g$. A subset $R \subset F(n,n)$ is said to be a min-representation of f if R is a finite set of (min, +) functions such that $f = \bigwedge_{g \in} g$. The mutual distributivity of $\otimes$ and $\wedge$ in the scalar case ($(x \oplus y) \wedge z = (x \wedge z) \oplus (y \wedge z)$ and $(x \wedge y) \oplus z = (x \oplus z) \wedge (y \oplus z)$) implies that every (min, max, +) function has a max-representation and a min-representation. Briefly, any time stream event graph for semantic rules "And" and "Weak-And" can be modeled under the following general form whose type is ((max, +), (min, max, +)):

$$\begin{cases} x(k) \geq f^{-}(x(k-1), x(k), u(k)) = A_1^{-} \otimes x(k-1) \\ \quad \oplus A_0^{-} \otimes x(k) \oplus B^{-} u(k) \text{ and} \\ x(k) \leq f^{+}(x(k-1), x(k), u(k)) = \bigwedge_{i=1}^{i_{\max}} g_i(x(k-1), \\ \quad x(k), u(k)) \text{ where} \\ g_i(x(k-1), x(k), u(k)) = A_{1,i}^{+} \otimes x(k-1) \\ \quad \oplus A_{0,i}^{+} \otimes x(k) \oplus B_i^{+} \otimes u(k) \end{cases} \quad [4.7]$$

The finite temporizations of $A_1^-, A_0^-, B^-, A_{1,i}^+, A_{0,i}^+$ and B_i^+ belong to $\mathbb{R}^+$ and $i_{\max}$ is the maximal number of (max,+) terms in function f^+.

The general form [4.7] includes the state equation of timed event graphs (equation [4.4]) because the relevant interval model is found by taking $i_{\max} = 1$. We deduce $A_1^- = A_1^+ = A_1$, $A_0^- = A_0^+ = A_0$ and $B^- = B^+ = B$. Consequently, the algebraic model of timed event graphs may be seen as a particular case of the interval model. Moreover, the case where each function $g_i(x(k-1), x(k), u(k))$ does not use the maximum operation defines a subclass that includes the model of P-time event graphs. Equally, let us note that the inequalities of P-time event graphs correspond to the semantic rule "And" of time stream event graphs.

REMARK.– Functions g_i in the upper term $f^+(x(k-1), x(k), u(k))$ must satisfy a non-degeneracy condition that we will describe in this section. The concatenation of $A^+_{1,i}$, $A^+_{0,i}$ and B^+_i is denoted by M_i and the notation $M_i(j,.)$ represents line j of matrix M_i, and $M_i(.,l)$ represents column l of matrix M_i. For every row j, $M_i(j,.)$ must be different from ε or, in other words, each row of M_i must describe at least one connection with an input or an incoming internal transition. Some elements $M_i(j,l)$ of row j in $A^+_{1,i}$ or $A^+_{0,i}$ naturally equal ε (respectively, in B^+_i) if there are no places connecting internal transition directly from x_l to x_j (from input transition u_l to internal transition x_j, respectively). If row $M_i(j,.)$ does not describe at least one connection, it becomes neutral by taking $M_i(j,.) = +\infty$.

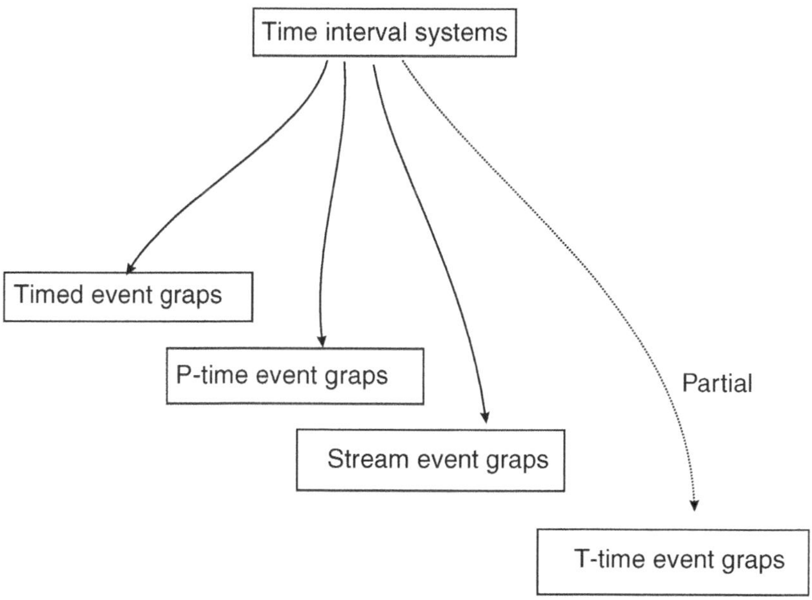

Figure 4.1. *Time interval systems*

Briefly, the interval model [4.9] whose type is ((max, +), (min, max, +)) can describe a large class of models composed of (at least) timed event graphs, P-time event graphs and time stream event graphs for the semantics "And" and "Weak-And". Note that the interval model cannot completely describe a T-time event graph as this last one contains a logical condition of the transition firing that cannot easily be rewritten (see the doctoral thesis by Abdelhak Guezzi [GUE 10]).

4.3. Control synthesis

In this chapter, the following assumption is made: for each interval model, the type of the lower bound denoted by f^- is (max, +). Therefore, $f^-(x(k-1), x(k), u(k)) = A_1^- \otimes x(k-1) \oplus A_0^- \otimes x(k) \oplus B^- u(k)$. Model [4.1] with this assumption will be considered below.

4.3.1. *Problem*

Let us assume that an interval model describes the specifications of a timed event graph. The first aim is the determination of an arbitrary trajectory following these models. Moreover, we may be given a sequence of dates at which we would like to see events occur at the latest, and we are asked to provide the latest input dates that would meet that objective. Therefore, the goal is to obtain the greatest control of a timed event graph when the state and control trajectories are constrained by additional specifications. This problem is the generalization of the classical just-in-time control of timed event graphs where the "backward" equations express the optimal control (section 5.6 in [BAC 92], [COC 93]). In other words, the "just-in-time" objective is to calculate the greatest control u such that $y \leq z$ with $y(k) = C \otimes x(k)$, knowing:

a) the trajectory of the desired output z;

b) the model of the plant described by a timed event graph:

$$x(k) = A_1 \otimes x(k-1) \oplus A_0 \otimes x(k) \oplus B \otimes u(k) \qquad [4.8]$$

c) the model of additional specifications expressed by an interval system [4.1]:

$$f^-(x(k-1), x(k), u(k)) \leq x(k) \leq f^+(x(k-1), x(k), u(k)) \quad [4.9]$$

where $f^-(x(k-1), x(k), u(k)) = A_1^- \otimes x(k-1) \oplus A_0^- \otimes x(k) \oplus B^- u(k)$ and f^+ is a (min, max, +) function.

The problem is defined on a horizon denoted by $[k_s, k_f]$ (where s represents "starts", and f represents "final").

4.3.2. *Pedagogical example: education system*

4.3.2.1. *Plant*

As shown in Figure 4.2, the plant corresponds to the following work of a professor: the lesson is composed of a lecture (duration: T_1) followed by practical work (duration: T_2). Transitions express the following events. x_1: beginning of the lecture; x_2: beginning of the practical work; x_3: end of the practical work; u_1: decision to start the lecture; and u_2: decision to start the practical work.

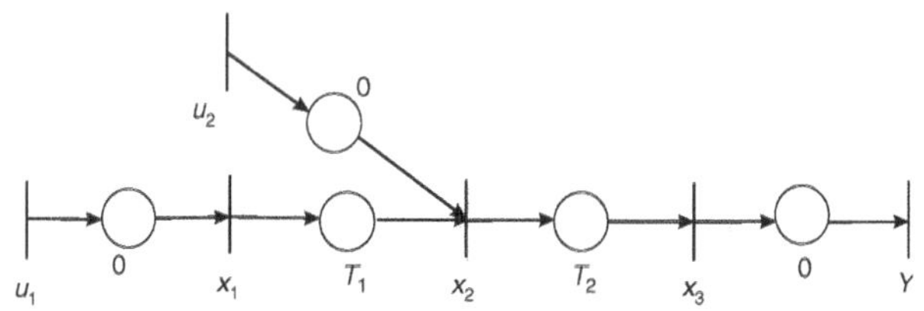

Figure 4.2. *Timed event graph (plant: tasks of the professor)*

Moreover, the professor must follow the official instructions: a lesson must not exceed T_3^+ and must not be less than T_3^-. These specifications can be described by a new event graph that can be a simple P-time event graph. So, $T_3^- \otimes x_1(k) \leq x_3(k) \leq T_3^+ \otimes x_1(k)$.

But, we can add more complex specifications. The first student is present at the lecture and the practical work (duration $[T_4^-, T_4^+]$). The second student is only present at the practical work (duration $[T_5^-, T_5^+]$). Moreover, the lesson stops with the departure of the last student: when the students switch off, they go out (specification 2). This new specification can be described by a time stream event graph (specification 2: two students)

$$T_4^- \otimes x_1(k) \oplus T_5^- \otimes x_2(k) \leq x_3(k) \leq T_4^+ \otimes x_1(k) \oplus T_5^+ \otimes x_2(k)$$

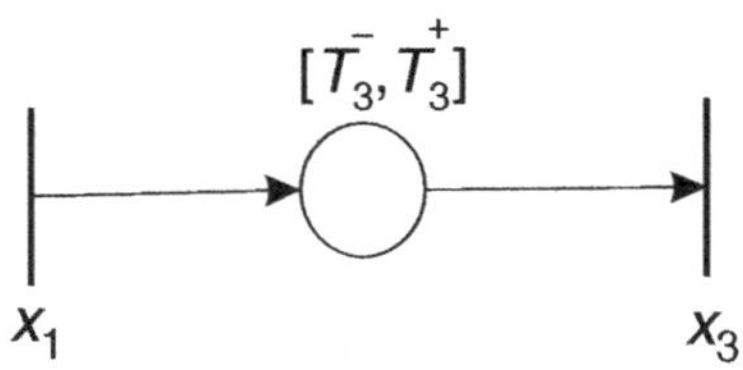

Figure 4.3. *P-time event graph (specification 1: official instructions)*

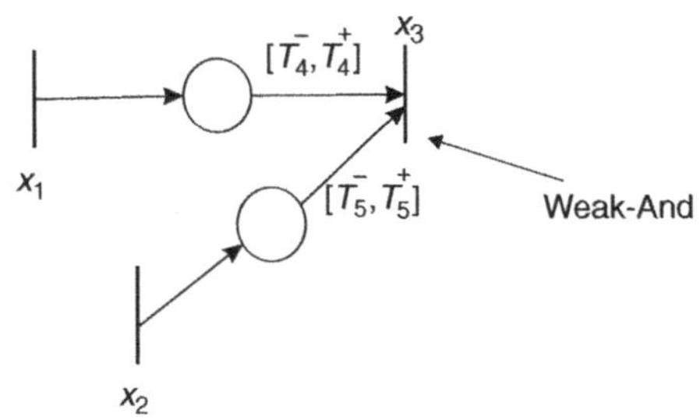

Figure 4.4. *Time stream event graph (specification 2: two students)*

4.3.2.2. *Problem*

The lesson must stop before the daily closing time of the school, which corresponds to the desired output z. The problem is the determination of the latest times to begin the lesson such that each specification is satisfied. If the teacher begins the lesson after this date, which corresponds to the greatest control u, the practical work cannot be finished ($y \nleq z$).

4.3.3. *Algebraic models*

Let us write the algebraic model of the different time event graphs.

$$x(k) = A_1 \otimes x(k-1) \oplus A_0 \otimes x(k) \oplus B \otimes u(k)$$
$$y(k) = C \otimes x(k)$$

$$A_1 = \varepsilon, A_0 = \begin{pmatrix} \varepsilon & \varepsilon & \varepsilon \\ T_1 & \varepsilon & \varepsilon \\ \varepsilon & T_2 & \varepsilon \end{pmatrix}, B = \begin{pmatrix} 0 & \varepsilon \\ \varepsilon & 0 \\ \varepsilon & \varepsilon \end{pmatrix} \text{ and}$$

$$C = \begin{pmatrix} \varepsilon & \varepsilon & 0 \end{pmatrix}$$

P-time event graph (specification 1: official instructions)

$$\begin{cases} x(k) \geq A_1^- \otimes x(k-1) \oplus A_0^- \otimes x(k) \oplus B^- u(k) \\ x(k) \leq f^+(x(k-1), x(k), u(k)) \end{cases}$$

$$A_1^- = \varepsilon, A_0^- = \begin{pmatrix} \varepsilon & \varepsilon & \varepsilon \\ \varepsilon & \varepsilon & \varepsilon \\ T_3^- & \varepsilon & \varepsilon \end{pmatrix}, B^- = \varepsilon \text{ and}$$

$$f^+(x(k-1), x(k), u(k)) = \begin{pmatrix} +\infty \\ +\infty \\ T_3^+ \otimes x_1(k) \end{pmatrix}$$

Time stream event graph (specification 2: two students)

$$\begin{cases} x(k) \geq A_1^- \otimes x(k-1) \oplus A_0^- \otimes x(k) \oplus B^- u(k) \\ x(k) \leq f^+(x(k-1), x(k), u(k)) \end{cases}$$

$$A_1^- = \varepsilon, A_0^- = \begin{pmatrix} \varepsilon & \varepsilon & \varepsilon \\ \varepsilon & \varepsilon & \varepsilon \\ T_4^- & T_5^- & \varepsilon \end{pmatrix}, B^- = \varepsilon \text{ and}$$

$$f^+(x(k-1), x(k), u(k)) = \begin{pmatrix} +\infty \\ +\infty \\ T_4^+ \otimes x_1(k) \oplus T_5^+ \otimes x_2(k) \end{pmatrix}$$

Interval system (specifications 1 and 2)

The following interval system describes specifications 1 and 2 and the relevant event graphs

$$\begin{cases} x(k) \geq A_1^- \otimes x(k-1) \oplus A_0^- \otimes x(k) \oplus B^- u(k) \\ x(k) \leq f^+(x(k-1), x(k), u(k)) \end{cases}$$

$$A_1^- = \varepsilon, A_0^- = \begin{pmatrix} \varepsilon & \varepsilon & \varepsilon \\ \varepsilon & \varepsilon & \varepsilon \\ T_3^- \oplus T_4^- & T_5^- & \varepsilon \end{pmatrix}, B^- = \varepsilon \text{ and}$$

$$f^+(x(k-1), x(k), u(k))$$

$$= \begin{pmatrix} +\infty \\ +\infty \\ T_3^+ \otimes x_1(k) \wedge T_4^+ \otimes x_1(k) \oplus T_5^+ \otimes x_2(k) \end{pmatrix}$$

which is a (min, max, +) function.

The following inequalities can be obtained from the different event graphs. This set corresponds to subsystem 1 in the following sections.

$$\begin{cases} x_1 \leq -T_1 + x_2 \\ x_2 \leq -\max(T_2, T_5^-) + x_3 \\ x_3 \leq +T_3^+ + x_1 \end{cases}$$

Therefore, $x_1 \leq -T_1 - \max(T_2, T_5^-) + T_3^+ + x_1$. If $T_1 = 60$, $T_2 = 90$, $T_3^+ = 175$ and $T_5^- = 0$, the inequality becomes $x_1 \leq -60 - \max(90, 0) + 175 + x_1 = 25 + x_1$ which is consistent. Now, if $T_3^+ = 145$, an incoherency appears as the inequality gives $x_1 \leq -5 + x_1$. The interpretation is that the lesson time of the professor is inconsistent with the official instructions. □

4.4. Fixed-point approach

4.4.1. *Fixed-point formulation*

The problem is now reformulated as a fixed-point problem.

THEOREM 4.1.– *For a timed event graph [4.8] following an interval model [4.9], the problem of the greatest control can be written as follows: from* k_s *to* k_f*, search the greatest state and control of the following inequality*

$$V \leq fh(V) \text{ with } V = \begin{pmatrix} x(k-1) \\ x(k) \\ x(k+1) \\ u(k) \end{pmatrix} \text{ and}$$

$$fh(V) = \begin{pmatrix} x(k-1) \\ fh_{bw}(x(k+1)) \wedge fh_{st}(x(k)) \wedge fh_{fw}(x(k-1), \\ \qquad x(k), u(k)) \\ x(k+1) \\ (B \oplus B^-)\backslash x(k) \end{pmatrix} \qquad [4.10]$$

where $fh_{bw}(x(k+1)) = (A_1 \oplus A_1^-)\backslash x(k+1)$*,* $fh_{st}(x(k)) = (A_0 \oplus A_0^-)\backslash x(k) \wedge (C\backslash z(k))$ *and* $fh_{fw}(x(k-1), x(k), u(k)) = [A_1 \otimes x(k-1) \oplus A_0 \otimes x(k) \oplus B \otimes u(k)] \wedge f^+(x(k-1), x(k), u(k))$.

PROOF.– The models are:

$$f^-(x(k-1), x(k), u(k)) \leq x(k) \leq f^+(x(k-1), x(k), u(k)),$$

where

$$f^-(x(k-1), x(k), u(k)) = A_1^- \otimes x(k-1) \oplus A_0^- \otimes x(k) \oplus B^- u(k)$$

and $f^+(x(k-1), x(k), u(k))$ is a (min, max, +) function.

$$x(k) = A_1 \otimes x(k-1) \oplus A_0 \otimes x(k) \oplus B \otimes u(k)$$

or

$$\begin{cases} A_1 \otimes x(k-1) \oplus A_0 \otimes x(k) \oplus B \otimes u(k) \leq x(k) \\ x(k) \leq A_1 \otimes x(k-1) \oplus A_0 \otimes x(k) \oplus B \otimes u(k) \end{cases}$$

Therefore, the problem can be written as

$$\begin{cases} (A_1 \oplus A_1^-) \otimes x(k-1) \oplus (A_0 \oplus A_0^-) \\ \qquad \otimes x(k) \oplus (B \oplus B^-) \otimes u(k) \leq x(k) \\ x(k) \leq [A_1 \otimes x(k-1) \oplus A_0 \otimes x(k) \\ \qquad \oplus B \otimes u(k)] \wedge f^+(x(k-1), x(k), u(k)) \\ y(k) = C \otimes x(k) \\ y(k) \leq z(k) \end{cases}$$

Consequently

$$\begin{cases} x(k) \leq (A_1 \oplus A_1^-)\backslash x(k+1) \wedge (A_0 \oplus A_0^-)\backslash x(k) \\ u(k) \leq (B \oplus B^-)\backslash x(k) \\ x(k) \leq [A_1 \otimes x(k-1) \oplus A_0 \otimes x(k) \oplus B \otimes u(k)] \\ \qquad \wedge f^+(x(k-1), x(k), u(k)) \\ x(k) \leq C\backslash z(k) \end{cases}$$

and the fixed-point formulation [4.10] is found. □

Expression [4.10] presents not only a "backward" part $fh_{bw}(x(k+1)$ but also a "static" part $fh_{st}(x(k), u(k))$ and a "forward" part $fh_{fw}(x(k-1), x(k), u(k))$. This fact increases the complexity of the problem and forbids the writing of simple equations on a short horizon such as the classical backward equations. Moreover, even if there is no function $f^+(x(k-1), x(k), u(k))$ in the specifications, fh is a (min, max, +) function. Briefly, we must solve a fixed-point problem of type $x \leq f(x)$ (if x exists) over the horizon of the desired output z with complex backward and forward interconnections defined by (min, max, +) functions.

EXAMPLE (education system continued).–

$$fh_{bw}(x(k+1)) = +\infty, fh_{st}(x(k))$$

$$= \begin{pmatrix} T_1\backslash x_2(k) \wedge (T_3^- \oplus T_4^-)\backslash x_3(k) \\ (T_2 \oplus T_5^-)\backslash x_3(k) \\ z(k) \end{pmatrix}$$

$$fh_{fw}(x(k-1), x(k), u(k))$$
$$= \begin{pmatrix} u_1(k) \\ T_1 \otimes x1(k) \oplus u_2(k) \\ T_2 \otimes x_2(k) \wedge T_3^+ \otimes x_1(k) \\ \wedge(T_4^+ \otimes x_1(k) \oplus T_5^+ \otimes x_2(k)) \end{pmatrix} \text{ and}$$

$$(B \oplus B^-)\backslash x(k) = \begin{pmatrix} x_1(k) \\ x_2(k) \end{pmatrix}.$$

Therefore

$$fh(V) = \begin{pmatrix} x(k-1) \\ T_1\backslash x_2(k) \wedge (T_3^- \oplus T_4^-)\backslash x_3(k) \wedge u_1(k) \\ (T_2 \oplus T_5^-)\backslash x_3(k) \wedge (T_1 \otimes x_1(k) \oplus u_2(k)) \\ z(k) \wedge T_2 \otimes x_2(k) \wedge T_3^+ \otimes x_1(k) \\ \qquad \wedge(T_4^+ \otimes x_1(k) \oplus T_5^+ \otimes x_2(k)) \\ x(k+1) \\ x_1(k) \\ x_2(k) \end{pmatrix} \qquad \square$$

4.4.2. *Existence*

The direct application of the famous Knaster–Tarski theorem will *a priori* define the solution set.

THEOREM 4.2.– *[KUM 95] [ZEI 86] Let* $(\mathcal{SX}, \leq)$ *be a complete lattice and* $f: \mathcal{SX} \to \mathcal{SX}$ *be a monotone function. Let* $\mathcal{SY} = \{x \in \mathcal{SX} \mid f(x) = x\}$ *be the set of fixed points of* f*. Then*

a) $\inf \mathcal{SY} \in \mathcal{SY}$*, and* $\inf \mathcal{SY} = \inf\{x \in \mathcal{SX} \mid f(x) \leq x\}$

b) $\sup \mathcal{SY} \in \mathcal{SY}$*, and* $\sup \mathcal{SY} = \sup\{x \in \mathcal{SX} \mid x \leq f(x)\}$

PROPOSITION 4.1.– Let $fh(V)\colon \overline{\mathbb{R}}_{\max}^{|V|} \to \overline{\mathbb{R}}_{\max}^{|V|}$ be defined by [4.10]. Let $\mathcal{SY} = \{V \in \overline{\mathbb{R}}_{\max}^{|V|} \mid fh(V) = V\}$ be the set of fixed points of f. Then:

a) $\inf \mathcal{SY} \in \mathcal{SY}$, and $\inf \mathcal{SY} = \inf\{V \in \overline{\mathbb{R}}_{\max}^{|V|} \mid fh\,(V) \leq V\}$

b) $\sup \mathcal{SY} \in \mathcal{SY}$, and $\sup \mathcal{SY} = \sup\{V \in \overline{\mathbb{R}}_{\max}^{|V|} \mid V \leq fh\,(V)\}$

PROOF.– The proof is immediate because $(\mathbb{R} \cup \{-\infty\} \cup \{+\infty\}, \leq)$ is a complete lattice and function $fh\,(\cdot)$ is isotone. If an algorithm gives the greatest solution of $x \leq f(x)$, this solution also satisfies the relevant equality. Function $fh(.)$ is also discontinuous but the Knaster–Tarski theorem does not require continuity of the function. □

Let us remark that the previous proposition does not lead to the existence of a finite solution because the considered lattice includes infinite elements. The aim of this part is to satisfy the existence of a finite solution in the control synthesis by using Theorem 4.3 presented below.

The property of homogeneity of functions (min, max, +) belonging to $F(n, m)$ is necessary to use the spectral vector below. This is defined as follows: $\forall \Delta \in \mathbb{R}, \forall x \in \mathbb{R}^n\ f(\Delta \otimes x) = \Delta \otimes f(x)$ in the usual vector–scalar convention: $(\Delta \otimes x)_i = \Delta \otimes x_i$.

As function fh contains the desired output z, fh is not a (min, max, +) function of type (n, m) that is homogeneous. The form of our practical problem is to find the greatest x (if x exists) such that $x \leq f(x)$ with f a non-homogeneous (min, max, +) function that can be defined by the following construction: $f = b, x_1, x_2, \ldots, x_n | f \otimes a | f \wedge f | f \oplus f$, where a, b are arbitrary real numbers ($a, b \in \mathbb{R}$). In the aim of applying the spectral theory about these functions, we will use a relaxation by associating a variable x_0 with b such that b is replaced by $b \otimes x_0$ in the above definition. So, the problem is to find the greatest $y = (x_0, x_1, ..., x_n)^t$ (if y exists) such that $x_0 = 0$ and $x_i \leq f_i(x_0, x_1, ..., x_n)^t)$ for $i \neq 0$. If we introduce the obvious inequality $x_0 \leq x_0$, the general problem is as follows: find the greatest $y = (x_0, x_1, ..., x_n)^t$ (if y exists) such that $x_0 \leq x_0$ and $x_i \leq f_i(x_0, x_1, ..., x_n)^t)$ for $i \neq 0$ with $x_0 = 0$. In other words, we have to solve the new system

$$y \leq f_{\text{hom}}(y)$$

where f_{hom} is a homogeneous function of type $(n+1, n+1)$.

Now, we remember some useful results of spectral theory. Dynamics of the form are considered: $x(k) = f(x(k-1))$, $\forall k \geq 1$ and $x(0) = \xi \in \mathbb{R}^n$, where f is a (min, max, +) function of type (n, n) $\mathbb{R}^n \to \mathbb{R}^n$. The cycle time vector is defined by $\chi(f) = \lim_{k\to\infty} x(k)/k$ if it exists. It does not depend on ξ. The pair $(\eta, v) \in \mathbb{R}^2$ is an ultimately affine regime of f if there exists an integer K such that $\forall k \geqslant K$, $f(v + k\eta) = v + (k+1)\eta$. In the following theorem, the notions of cycle time and ultimately affine regime, which always exist in $F(n, n)$, make it possible to satisfy the existence of a solution of different inequalities and equalities. The proof of the following fundamental theorem gives a complete presentation of the approach.

THEOREM 4.3.– *[GAU 98] [COC 99] [COC 01] [DID 05] Let f be a function of $F(n, n)$. The two following conditions are equivalent:*

i) there is a finite x such that $x \leq f(x)$ (respectively, $x \geq f(x)$);

ii) $\chi(f) \geq 0$ (respectively, $\chi(f) \leq 0$).

PROOF.– (i)⇒(ii) $x \leqslant f(x) \Rightarrow x \leqslant f(x) \leqslant f^2(x) \leqslant f^k(x)$ with $k \geqslant 1$ because f is non-decreasing. By definition, $\chi(f, x) = \lim_{k\to\infty} x(k)/k = \lim_{k\to\infty} f^k(x)/k \geqslant \lim_{k\to\infty} x/k = 0$ (x is finite).

(ii)⇒(i) In [COC 01], it has been proved that any function in $F(n, n)$ has an ultimately affine regime and, moreover, $\chi(f)$

equals η, for all ultimately affine regimes $(\eta, v) \in (R^n)^2$ of f. For an ultimately affine regime (η, v), we have $\forall k \geqslant K$, $f(v + k\eta) = v + (k+1)\eta$. We put $y = v + k\eta$ with $k \geqslant K$. The variable y satisfies $f(y) = y + \eta$. As $\eta = \chi(f) \geqslant 0$, we deduce that there exists y such that $f(y) \geqslant y$. □

NOTATION.– Let us denote $X_l = (x(k)^t, x(k+1)^t, ..., x(k+l)^t)^t$ with horizon l and the dater algebraic function $\Phi_f(X_l)$ obtained by developing $f(x(k))$ algebraically on X_l. Therefore, $X_l \leq \Phi_f(X_l)$ describes every relation between the components of X_l. The following example illustrates this notation.

EXAMPLE OF NOTATION.–

Let $f(x(k)) = \binom{a \otimes x_1(k) \oplus b \otimes x_2(k+1)}{c \otimes x_2(k) \oplus d \otimes x_1(k-1)}$ with $x(k) = \binom{x_1(k)}{x_2(k)}$.

$X_2 = (x_1(k), x_2(k), x_1(k+1), x_2(k+1), x_1(k+2), x_2(k+2))^t$

$\Phi_f(X_2) = (a \otimes x_1(k) \oplus b \otimes x_2(k+1), c \otimes x_2(k), a \otimes x_1(k+1) \oplus b \otimes x_2(k+2), c \otimes x_2(k+1) \oplus d \otimes x_1(k), a \otimes x_1(k+2), c \otimes x_2(k+2) \oplus d \otimes x_1(k+1))^t$. □

Below, function fh_{hom} is deduced from fh defined in relation [4.10].

THEOREM 4.4.– *Let us consider*

$$\begin{pmatrix} x_0 \\ V \end{pmatrix} \leq fh_{\text{hom}} \begin{pmatrix} x_0 \\ V \end{pmatrix} \quad [4.11]$$

where

$$fh_{\text{hom}} \begin{pmatrix} x_0 \\ V \end{pmatrix} = \begin{pmatrix} x_0 \\ x(k-1) \\ fh_{bw}(x(k+1)) \wedge fh_{st}(x_0, x(k)) \\ \qquad \wedge fh_{fw}(x(k-1), x(k), u(k)) \\ x(k+1) \\ (B \oplus B^-) \backslash x(k) \end{pmatrix}$$

and $X_l = (x_0, x(k)^t, u(k)^t, x(k+1)^t, u(k+1)^t, ..., x(k+l)^t, u(k+l)^t)^t$.

There exists a finite solution satisfying system [4.7] on horizon l *if and only if* $\chi(\Phi_{fh_{\text{hom}}}(X_l)) \geq 0$.

PROOF.– In $x(k) \leq fh_{bw}(x(k+1)) \wedge fh_{st}(x(k)) \wedge fh_{fw}(x(k-1), x(k), u(k))$, constraint $x(k) \leq fh_{st}(x(k)) = (A_0 \oplus A_0^-)\backslash x(k) \wedge (C\backslash z(k))$ becomes $x(k) \leq fh_{st}(x_0, x(k)) = (A_0 \oplus A_0^-)\backslash x(k) \wedge (C\backslash(z(k) \otimes x_0))$. Consequently, the system defined by [4.10] can also be defined by system [4.11] under the condition $x_0 = 0$. The final inequality set presents the general form $x \leq f_{\text{hom}}(x)$ that can be associated with the algebraic inequality $X_l \leq \Phi_F(X_l)$. As fh_{hom} is a homogeneous function, the spectral vector can be calculated $\chi(\Phi_{fh_{\text{hom}}}(X_l))$ and Theorem 4.3 can be applied. If the cycle time satisfies the corresponding condition of existence, the interval system presents a solution. For any solution, an obvious translation can be applied in such a way that the first component equals zero. The existence of a solution implies the possibility of taking $x_0 = 0$ because of the homogeneity of fh_{hom}. □

REMARK.– Let us note that product $z(k) \otimes x_0$ can also be written as $x_0 \otimes z(k)$.

Before giving an example, the following preliminary results give a way of calculating the spectral vector.

Preliminary results

The calculation of the spectral vector can be done as follows. Remember that an associated graph (or an induced graph) of a square matrix A is denoted by $G(A)$ and is deduced from this matrix by associating a node i with column i and line

i. For each $A_{ij} \neq \varepsilon$, an arc from node j toward node i is added. The following convention will be useful to (min, max, +) functions: all arcs derived from the same (max, +) row are bundled together with a line drawn through them near their arrow end [WAL 95] (see Figures 4.5 and 4.7). Also remember that the weight of a path pa, $\mid pa \mid_w$ is the sum of the labels on the edges of the path. The length of a path pa, $\mid pa \mid_l$ is the number of edges of the path. A circuit is a path that starts from and ends at the same node. If c is a circuit, its cycle average is defined by $\mid c \mid_w / \mid c \mid_l$ (the notation / represents the classical division). A node j is upstream from i, denoted $i \Longleftarrow j$, if either $i = j$ or if there is a path in $G(A)$ from j to i. A circuit c is upstream from i, denoted $i \Longleftarrow c$, if either $i \in c$ or if there is a path in $G(A)$ from a node $j \in c$ to i. Vector $\mu(A) \in \mathbb{R}^n$ is defined by $\mu_i(A) = \max\{\mid c \mid_w / \mid c \mid_l \mid i \Longleftarrow c\}$. If $f \in F(n,n)$ is (max, +) and A is the associated matrix over $\mathbb{R}_{\max}$, then $\chi(f)$ exists and $\chi(f) = \mu(A)$. The result is identical for a (min, +) function.

Let $f \in F(n,n)$. Remember that a subset $S \subset F(n,n)$ is said to be a max-representation of f if S is a finite set of (max, +) functions such that $f = \bigwedge_{g \in S} g$. As each function of $F(n,n)$ is made up of n functions belonging to $F(n,1)$, the set of (min, max, +) functions $F(n,n)$ has a natural representation as an n-fold Cartesian product: $F(n,n) = F(n,1) \times ... F(n,1)$. Let S_i be set $\{fh_i$ such that $fh_i \in S\}$. The rectangularization of S, denoted $rec(S)$, is defined by $rec(S) = S_1 \times S_2 \times ... \times S_n$. In other words, a set S of (min, max, +) functions is rectangular if for all fh, $fh' \in S$, and for all $i = 1, ..., n$, the function obtained by replacing the i-th component of fh by the ith component of fh' belongs to S. Therefore, $rec(S)$ is finite when S is finite and $S \subset rec(S)$. If S and R are rectangular max- and min-representations, respectively, of $f \in F(n,n)$, then $\chi(f) = \bigwedge_{fh \in S} \chi(fh) = \bigoplus_{g \in R} \chi(g)$.

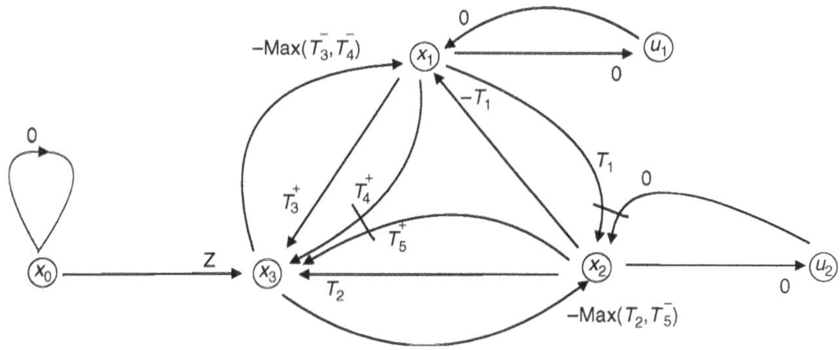

Figure 4.5. *Associated graph of the pedagogical example*

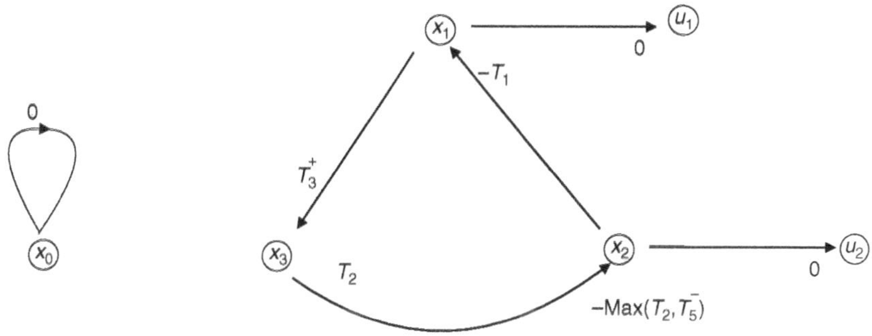

Figure 4.6. *Subsystem 1*

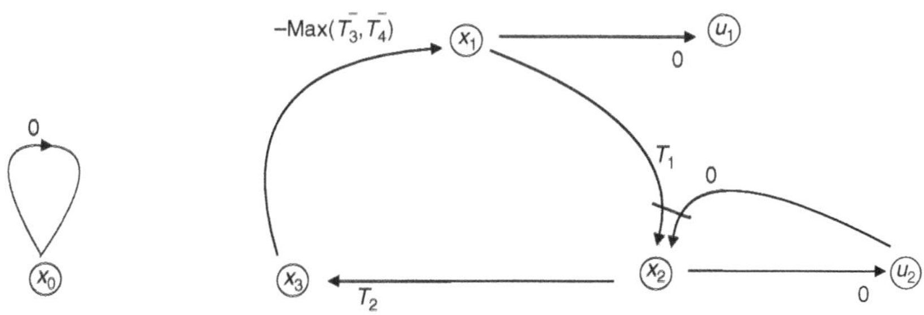

Figure 4.7. *Subsystem 2*

Example continued (education system).–

$$rec(S) = \{L_{11}\} \times \{L_{21}\} \times \{L_{31}\} \times \{L_{41}\} \times \{L_{51}, L_{52}, L_{53}\}$$
$$\times \{L_{61}, L_{62}\} \times \{L_{71}, L_{72}, L_{73}, L_{74}\} \times \{L_{81}\}$$

$L_{11}(V) = x_0;$

$L_{21}(V) = x_1(k-1);$

$L_{31}(V) = x_2(k-1);$

$L_{41}(V) = x_3(k-1);$

$L_{51}(V) = T_1 \backslash x_2(k)$; $L_{52}(V) = (T_3^- \oplus T_4^-)\backslash x_3(k)$; $L_{53}(V) = u_1(k);$

$L_{61}(V) = (T_2 \oplus T_5^-)\backslash x_3(k)$; $L_{62}(V) = T_1 \otimes x_1(k) \oplus u_2(k);$

$L_{71}(V) = z(k) \otimes x_0$; $L_{72}(V) = T_2 \otimes x_2(k)$; $L_{73}(V) = T_3^+ \otimes x_1(k)$; $L_{74}(V) = (T_4^+ \otimes x_1(k) \oplus T_5^+ \otimes x_2(k));$

$L_{81}(V) = x_1(k+1);$

$L_{91}(V) = x_2(k+1);$

$L_{10,1}(V) = x_3(k+1);$

$L_{11,1}(V) = x_1(k);$

$L_{12,1}(V) = x_2(k).$ □

The (max, +) subsystems ($3 \cdot 2 \cdot 4 = 24$ possibilities) are produced by a combination of the previous functions. Their analysis leads to $\chi(\Phi_{fh_{\text{hom}}}(X_{+\infty})) = 0$ for the following values: $T_1 = 60$; $T_2 = 90$; $T_3^- = 165$; $T_3^+ = 175$; $T_4^- = 0$; $T_4^+ = 143$; $T_5^- = 0$; and $T_5^+ = 98$. Below, the calculation is considered in two examples of subsystems. As the process is "static", the functions $L_{21}(V), L_{31}(V), L_{41}(V), L_{81}(V), L_{91}(V)$ and $L_{10,1}(V)$ can be disregarded.

Subsystem 1

$$\begin{pmatrix} L_{11}(V) = x_0 \\ L_{51}(V) = T_1 \backslash x_2(k) \\ L_{61}(V) = (T_2 \oplus T_5^-) \backslash x_3(k) \\ L_{73}(V) = T_3^+ \otimes x_1(k) \\ L_{11,1}(V) = x_1(k) \\ L_{12,1}(V) = x_2(k) \end{pmatrix}$$

The substructure contains a unique circuit that must be positive or null. If $-T_1 - (T_2 \oplus T_5^-) + T_3^+ \geq 0$, the relevant spectral vector is positive or null.

Subsystem 2

$$\begin{pmatrix} L_{11}(V) = x_0 \\ L_{52}(V) = (T_3^- \oplus T_4^-) \backslash x_3(k) \\ L_{62}(V) = T_1 \otimes x_1(k) \oplus u_2(k) \\ L_{72}(V) = T_2 \otimes x_2(k) \\ L_{11,1}(V) = x_1(k) \\ L_{12,1}(V) = x_2(k) \end{pmatrix}$$

Functions L_{62} and $L_{12,1}$ constitute a null circuit and there is a path from x_2 to the other vertices. Even if the other circuit satisfies $-(T_3^- \oplus T_4^-) + T_1 + T_2 < 0$, the relevant spectral vector is null because the maximum vector is taken.

4.4.3. *Structure*

Now, the underlined structure of fh_{hom} will be considered. Function $fh_{\text{hom},1}$ defined below is similar to classical "backward equations" for timed event graphs.

$$\begin{pmatrix} x_0 \\ V \end{pmatrix} \leq fh_{\text{hom},1}\begin{pmatrix} x_0 \\ V \end{pmatrix} \text{ with } fh_{\text{hom},1}\begin{pmatrix} x_0 \\ V \end{pmatrix}$$

$$= \begin{pmatrix} x_0 \\ x(k-1) \\ A_1\backslash x(k+1) \wedge A_0\backslash x(k) \wedge (C\backslash(z(k) \otimes x_0)) \\ x(k+1) \\ B\backslash x(k) \end{pmatrix}$$

The structural observability (respectively, the controllability) gives a condition to observe an effect in the output (respectively, transition) whose origin comes from one internal transition (respectively, input) at least. It allows the introduction of the following propositions where an infinite horizon is considered.

DEFINITION 4.4.– *[BAC 92] An event graph is structurally controllable if every internal transition can be reached by a path from one input transition at least. An event graph is structurally observable if, from every internal transition, there exists a path to at least one output transition.*

PROPOSITION 4.2.– A structurally observable event graph defined by model [4.8] satisfies $\chi(\Phi_{fh_{\text{hom},1}}(X_{+\infty})) = 0$. Moreover, the greatest solution, denoted θ', of the "backward" part $fh_1(V) = V$ satisfies $fh(\theta') \leq \theta'$. The greatest solution of the "backward" part gives a finite upper bound of the solutions of the inequality [4.10] $V \leq fh(V)$ and the corresponding equality $V = fh(V)$.

PROOF.– The max-representation of $fh(V)$ is $\bigwedge_{i=1}^{+\infty} fh_i(V)$ with fh_i a (max, +) function. The corresponding homogeneous function is $fh_{\text{hom}}\begin{pmatrix} x_0 \\ V \end{pmatrix} = \bigwedge_{i=1}^{+\infty} fh_{\text{hom},i}\begin{pmatrix} x_0 \\ V \end{pmatrix}$.

A circuit whose weight equals zero is associated with x_0 because $x_0 \leq x_0$. Every known desired output value $z(k)$ defines an arc from vertex x_0 to a vertex associated with a component of the state for a given k. If we assume that the system is structurally observable, from every internal transition, there exists a path to one output transition in the event graph at least. In the associated graph, the direction of the paths is opposite, and for every vertex corresponding to a transition at any k, there is a path going from x_0. Moreover, the same reasoning can be made for the inputs: as every input transition is connected to the system by definition, each vertex associated with an input for k can be reached by a path from one internal transition at least. Therefore, the definition of the spectral vector shows that $\chi(\Phi_{fh_{\text{hom},1}}(X_{+\infty})) = 0$. Consequently, there is θ such that $\theta = fh_{\text{hom},1}(\theta)$. This solution also satisfies $fh_{\text{hom}}(\theta) = \bigwedge_{i=1}^{+\infty} fh_{\text{hom}\,i}(\theta) \leq fh_{\text{hom},1}(\theta) = \theta$. Particularly, the corresponding greatest solution of the "backward" part $fh_1(V) = V$ for $x_0 = 0$, denoted θ', satisfies $fh(\theta') \leq \theta'$. □

As the set of series is a complete lattice and fh_1 is a monotone function, Theorem 4.2 can be applied: it proves that the greatest solution θ' of $fh_1(V) = V$ is also the greatest solution satisfying

$$V \leq fh_1(V) \qquad [4.12]$$

This result can also be proved by considering the set defined by [4.12] that is a complete lattice (Theorem 4.72 point 2 in [BAC 92]). Moreover, as [4.10] $V \leq fh(V)$ is equivalent to the set of inequalities $V \leq fh_i(V)$, each variable V must satisfy each inequality and particularly [4.12]. Consequently, each solution V of [4.10] is less than or equal to θ'. Finally, this set $\{V \mid V \leq fh(V)\}$ includes the solution set satisfying $V = fh(V)$. The greatest solution of $V \leq fh(V)\}$is denoted by θ. □

REMARK.– Let us note that the reasoning does not need the assumption of structurally controllable system but only the definition of the input transition.

Example (education system continued).–

As the system is "static", we take $V = \begin{pmatrix} x(k) \\ u(k) \end{pmatrix}$.

$$A_1 \backslash x(k+1) \wedge A_0 \backslash x(k) \wedge (C \backslash (z(k))) = \begin{pmatrix} T_1 \backslash x_2(k) \\ T_2 \backslash x_3(k) \\ z(k) \end{pmatrix} \text{ and}$$

$B \backslash x(k) = \begin{pmatrix} x_1(k) \\ x_2(k) \end{pmatrix}$. If $z = 200$, $\theta' = \begin{pmatrix} 50 \\ 110 \\ 200 \\ 50 \\ 110 \end{pmatrix}$ for $fh_1(x(k)) = \begin{pmatrix} T_1 \backslash x_2(k) \\ T_2 \backslash x_3(k) \\ z(k) \\ x_1(k) \\ x_2(k) \end{pmatrix}$ and, using a classical fixed-point approach (see the following section 4.5), $\theta = \begin{pmatrix} 35 \\ 110 \\ 200 \\ 35 \\ 110 \end{pmatrix}$ for $fh_{bw}(x(k+1)) \wedge fh_{st}(x(k)) \wedge fh_{fw}(x(k-1), x(k), u(k)) =$

$$\begin{pmatrix} T_1 \backslash x_2(k) \wedge (T_3^- \oplus T_4^-) \backslash x_3(k) \wedge u_1(k) \\ (T_2 \oplus T_5^-) \backslash x_3(k) \wedge (T_1 \otimes x_1(k) \oplus u_2(k)) \\ z(k) \wedge T_2 \otimes x_2(k) \wedge T_3^+ \otimes x_1(k) \\ \wedge (T_4^+ \otimes x_1(k) \oplus T_5^+ \otimes x_2(k)) \end{pmatrix}.$$

Therefore, the greatest control u is $\begin{pmatrix} 35 \\ 110 \end{pmatrix}$. The calculations give $fh_1(\theta') = \theta'$; $fh(\theta') \leq \theta'$; $\theta = fh(\theta)$; and $\theta \leq \theta'$. □

As a structurally observable event graph defined by model [4.8] satisfies $\chi(\Phi_{fh_{\text{hom},1}}(X_{+\infty})) = 0$ (Proposition 4.5), we can deduce that the spectral vector of the complete system is less than or equal to zero. The association with new terms can create strictly negative circuits that can imply that there is no solution.

Example (education system continued).– As there is a path from node x_0 associated with a null loop ($x_0 \leq L_{11}(V) = x_0$) to each node, $\chi(\Phi_{fh_{\text{hom},1}}(X_{+\infty})) = 0$. Now, if $T_3^+ = 145$, $-T_1 - (T_2 \oplus T_5^-) + T_3^+ = -5 \geq 0$ and some components of spectral vector $\chi(\Phi_{fh_{\text{hom}}}(X_{+\infty}))$ are negative. Therefore, $\chi(\Phi_{fh_{\text{hom}}}(X_{+\infty})) < 0$ and [4.10] $V \leq fh\,(V)$ has no solution: the official instructions are too strict. □

4.5. Algorithm

The calculation of the greatest control of inequality [4.10] can be made by a classical iterative algorithm of McMillan and Dill [MCM 92] or Walkup-Borriello [WAL 95] (see also [HAS 99]) that particularizes the algorithm of Kleene to (min, max, +) expressions. The general resolution of $x \leq f(x)$ is given by the iterations of $x_{i+1} \leftarrow f(x_i) \wedge x_i$ if the finite starting point is greater than the final solution. Here, i represents the number of iterations and not the number of components of vector x. Following this framework, we provide an algorithm specific to the determination of the greatest state and control.

ALGORITHM 1.– Step 0 (initialization): $\rho(k) \leftarrow +\infty$ for $k \geq k_s$; $\delta(k_s) \leftarrow x_{k_s}$ and $\delta(k) \leftarrow +\infty$ for $k \geq k_s + 1$;

Repeat

– Step 1: for $k = k_f$ to k_s, $\rho(k) \leftarrow (A_0 \oplus A_0^-)^* \backslash [\delta(k) \wedge [(A_1 \oplus A_1^-) \backslash \rho(k+1)] \wedge [C \backslash z(k)]]$

– Step 2: for $k = k_f$ to k_s, $u(k) \leftarrow (B \oplus B^-)\backslash\rho(k)$

– Step 3: $\delta(k_s) \leftarrow \rho(k_s)$

For $k = k_s + 1$ to k_f, repeat $\delta(k) \leftarrow \rho(k) \wedge [A_1 \otimes \delta(k-1) \oplus A_0 \otimes \delta(k) \oplus B \otimes u(k)] \wedge f^+(\delta(k-1), \delta(k), u(k))$ until no $\delta_i(k)$ changes.

Until $\rho(k) = \delta(k)$ for $k_s \leq k \leq k_f$ or $\delta(k_s) \neq x_{k_s}$

The introduction of the variables ρ and δ allows the memorization of the past values of the state in the algorithm. The "backward" part corresponds to steps 1 and 2, and the first iteration allows the determination of the starting state trajectory ρ of the general algorithm before an optimal minimization. As variables $\delta(k)$ and $\rho(k)$ are contained in the right-hand terms of the expressions, step 1 results from the application of Theorem 4.73 part 1 in [BAC 92] to $\rho(k) \leftarrow \delta(k) \wedge [(A_1 \oplus A_1^-)\backslash\rho(k+1) \wedge (A_0 \oplus A_0^-)\backslash\rho(k)] \wedge [C\backslash z(k)]$ and step 3 improves the state value expressed by δ with a fixed-point approach applied to the relevant relation. Finally, when the minimization of the state stops, the algorithm gives the optimal state and control that satisfies the inequalities of the model. For instance, the algorithm gives a state trajectory of a timed event graph that satisfies the state equation. As the control is directly calculated from the state that is minimized at each step until the convergence, the minimization of the control does not need a memorization of their previous calculated values as for the state.

As the control is applied to a timed event graph whose evolution must start from an initial state, an arbitrary initial condition x_{k_s} is assumed to be known: this "non-canonical" initial condition can be the result of the past evolution of a process, for instance. Therefore, expression $x(k_s) = x_{k_s}$ is a new additional constraint that must be considered in the determination of the control. This constraint is introduced in initialization $\delta(k_s) \leftarrow x_{k_s}$ that corresponds to the relaxed form $x(k_s) \leq x_{k_s}$ because the algorithm calculates the greatest state trajectory and minimizes the initial values. Therefore, the solution presents two cases:

– The estimate of the greatest state trajectory satisfies $x^+(k_s) = x_{k_s}$ and therefore, there are a control and state trajectories that satisfy all constraints.

– The estimate of the greatest state trajectory satisfies $x(k_s) \neq x_{k_s}$, the previous problem is without solution. If the problem converges to a finite solution with $x(k_s) \leq x_{k_s}$ and $x(k_s) \neq x_{k_s}$, this case can be interpreted as a past evolution of the process whose trajectory is "too early" relatively to x_{k_s}.

Example (education system continued).– Using the algorithm, the greatest control u is $\begin{pmatrix} 35 \\ 110 \end{pmatrix}$ for $z = 200$. Also, $x = \begin{pmatrix} 35 \\ 110 \\ 200 \end{pmatrix}$. □

The development of the algorithm is easy and requires only the memorization of the matrices of the different models and the estimated trajectory. Generally, it is difficult to carry out a theoretical analysis of the number of iterations like in many algorithms in this field [CHE 05]. The general algorithm of McMillan and Dill [MCM 92] is known to be pseudo-polynomial [MCM 92] [WAL 95] in practice. The following result, however, shows that the proposed algorithm is very quick when there is no specification: the algorithm converges to the greatest state and control in one iteration. The model of the timed event graph is supposed to be $x(k) = A \otimes x(k-1) \oplus$

$B \otimes u(k)$. In this case, $A_1 \oplus A_1^- = A \oplus A_1^- = A$, $B \oplus B^- = B$ and $f^+(\delta(k-1), \delta(k), u(k)) = +\infty$. The following proposition proves that $\delta(k) \leq [A \backslash \delta(k+1)] \wedge [C \backslash z(k)]$, which implies that any new iteration of the loop as in the general algorithm will be useless.

PROPOSITION 4.3.– For an optimal control synthesis of a timed event graph without specification, the following inequality is satisfied at the end of step 2 of the algorithm:

$$\delta(k) \leq [(A \oplus A^-) \backslash \delta(k+1)] \wedge [C \backslash z(k)]$$

PROOF.– Simple substitutions and property of distributivity of the left-residual $\backslash$ relative to $\wedge$ are applied below:

$$\begin{aligned} & A \backslash \delta(k+1) \wedge C \backslash z(k) \\ & \quad = A \backslash [\rho(k+1) \wedge [A\mu(k) \oplus Bu(k+1)]] \wedge C \backslash z(k) \\ & \quad = A \backslash \rho(k+1) \wedge A \backslash [A\mu(k) \oplus Bu(k+1)] \wedge C \backslash z(k) \\ & \quad = A \backslash \rho(k+1) \wedge C \backslash z(k) \wedge A \backslash [A\mu(k) \oplus Bu(k+1)]. \end{aligned}$$

So,

$$A \backslash \delta(k+1) \wedge C \backslash z(k) = A \backslash \rho(k+1) \wedge C \backslash z(k) \wedge A \backslash [A\mu(k) \oplus Bu(k+1)].$$

– Let us consider the term $A \backslash \rho(k+1) \wedge C \backslash z(k)$ in the previous relation. $\rho(k) = A \backslash \rho(k+1) \wedge C \backslash z(k)$ at the first iteration because $\delta(k) = +\infty$ on horizon $[k_s + 1, k_f]$. However, $\rho(k_s) = \delta(k_s) \wedge [A \backslash \rho(k_s+1)] \wedge [C \backslash z(k_s)]$ and consequently, $A \backslash \rho(k+1) \wedge C \backslash z(k) \geq \rho(k)$ for k belonging to $[k_s, k_f]$.

Therefore, $A \backslash \delta(k+1) \wedge C \backslash z(k) \geq \rho(k) \wedge A \backslash [A\mu(k) \oplus Bu(k+1)]$.

– Now, consider $A \backslash [A\mu(k) \oplus Bu(k+1)]$. If f is residuated, then there exists fh such that $foh \leq Id$ and $hof \geq Id$ (see preliminaries on residuation in Chapter 2 (section 2.2) on consistency) and $A \otimes (A \backslash x) \leq Id$ and $A \backslash [A \otimes x] \geq Id$

Consequently, $A\backslash[A \otimes \delta(k)] \geq \delta(k)$ and $A\backslash[A \otimes \delta(k) \oplus B \otimes u(k+1)] \geq \delta(k)$ by isotony of the left residual.

– Moreover, by construction, step 3 implies $\rho(k) \geq \delta(k)$.

Briefly, $A\backslash\delta(k+1) \wedge C\backslash z(k) \geq \rho(k) \wedge A\backslash[A\mu(k) \oplus Bu(k+1)] \geq \delta(k)$, which terminates the proof. □

The previous result shows that the proposed approach generalizes the classical "backward" approach for timed event graphs. Let us remember that, in this case, the "backward" recursive equations explicitly give the greatest solution (the latest times) of the control problem where the covector plays the role of the state vector, whereas dater equations give the least solution (the earliest times) of the process evolution [BAC 92].

4.6. Example

The following example gives another illustration of the theoretical results. Computation tests are made using Maxplus toolboxes in Scilab. A simple example of time stream Petri nets is given in Figure 4.9 and describes the specifications of the timed event graph of Figure 4.8.

4.6.1. *Models*

4.6.1.1. *Plant: timed event graph*

$$x(k) = \begin{pmatrix} x_1(k) \\ x_2(k) \end{pmatrix}, u(k) = \begin{pmatrix} u_1(k) \\ u_2(k) \end{pmatrix}$$

$x(k) = A_1 \otimes x(k-1) \oplus B \otimes u(k)$ and $y(k) = C \otimes x(k)$, where $A_1 = \begin{pmatrix} \varepsilon & T_9 \\ T_8 & \varepsilon \end{pmatrix}$, $B = \begin{pmatrix} T_6 & \varepsilon \\ \varepsilon & T_7^- \end{pmatrix}$ and $C = (\varepsilon \;\; e)$.

$T_6 = 6$, $T_7 = 1$, $T_8 = 3$ and $T_9 = 2$.

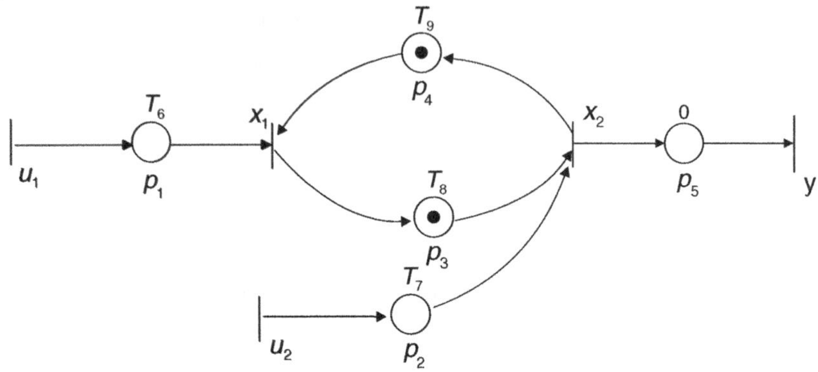

Figure 4.8. *Timed event graph*

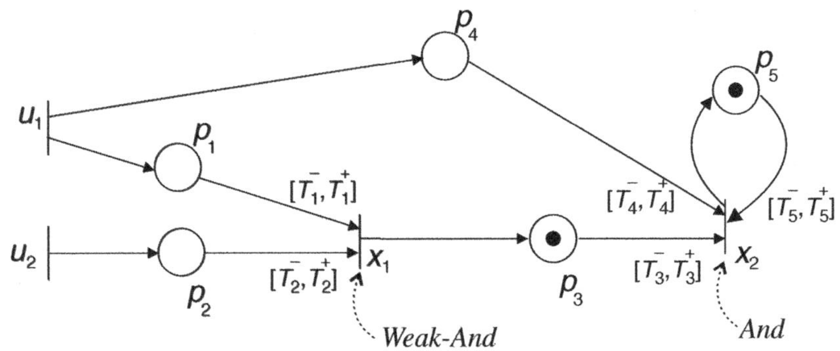

Figure 4.9. *Time stream event graph*

4.6.1.2. *Specifications: time stream event graph*

$x(k) \geq f^{-}(x(k-1), u(k)) = A_1^{-} \otimes x(k-1) \oplus B^{-}u(k)$, where

$$A_1^{-} = \begin{pmatrix} \varepsilon & \varepsilon \\ T_3^{-} & T_5^{-} \end{pmatrix}, B^{-} = \begin{pmatrix} T_1^{-} & T_2^{-} \\ T_4^{-} & \varepsilon \end{pmatrix}.$$

$$x(k) \leq f^{+}(x(k-1), u(k)) = \bigwedge_{i=1}^{i_{\max}} (A_i^{+} \otimes x(k-1) \oplus B_i^{+} \otimes u(k)),$$

where

$i_{\max} = 3$, $A_1^+ = \begin{pmatrix} \varepsilon & \varepsilon \\ T_3^+ & \varepsilon \end{pmatrix}$, $B_1^+ = \begin{pmatrix} T_1^+ & T_2^+ \\ \varepsilon & \varepsilon \end{pmatrix}$, $A_2^+ = \begin{pmatrix} +\infty & +\infty \\ \varepsilon & \varepsilon \end{pmatrix}$, $B_2^+ = \begin{pmatrix} +\infty & +\infty \\ T_4^+ & \varepsilon \end{pmatrix}$, $A_3^+ = \begin{pmatrix} +\infty & +\infty \\ \varepsilon & T_5^+ \end{pmatrix}$, $B_3^+ = \begin{pmatrix} +\infty & +\infty \\ \varepsilon & \varepsilon \end{pmatrix}$.

$[T_1^-, T_1^+] = [1,5]$; $[T_2^-, T_2^+] = [2,6]$; $[T_3^-, T_3^+] = [2,11]$; $[T_4^-, T_4^+] = [3,10]$; $[T_5^-, T_5^+] = [0,3]$.

4.6.2. *Fixed-point formulation*

The homogeneous form of [4.10] in Theorem 4.1 is defined by $fh_{\text{hom}}\begin{pmatrix} x_0 \\ V \end{pmatrix} = fh_{\text{hom},1}\begin{pmatrix} x_0 \\ V \end{pmatrix} \wedge fh_{\text{hom}\,2}\begin{pmatrix} x(k-1) \\ u(k) \end{pmatrix}$, where $fh_{\text{hom},1}$ and $fh_{\text{hom}\,2}$ are defined as follows:

$$fh_{\text{hom},1}\begin{pmatrix} x_0 \\ x(k-1) \\ x(k) \\ x(k+1) \\ u(k) \end{pmatrix} = \begin{pmatrix} x_0 \\ x(k-1) \\ (A_1 \oplus A_1^-)\backslash x(k+1) \\ \quad \wedge (C\backslash(z(k) \otimes x_0)) \\ x(k+1) \\ (B \oplus B^-)\backslash x(k) \end{pmatrix}$$

$$\begin{pmatrix} x_0 \\ (A_1 \oplus A_1^-)\backslash x(k+1) \wedge (C\backslash(z(k) \otimes x_0)) \\ (B \oplus B^-)\backslash x(k) \end{pmatrix} = \begin{pmatrix} x_0 \\ (T_3^- \oplus T_8) \backslash (x_2(k+1)) \\ (T_9\backslash(x_1 k^{-1})) \\ \wedge(T_5^-\backslash(x_2(k+1))) \wedge (z \otimes x_0) \\ ((T_6 \oplus T_1^-) \backslash x_1) \wedge (T_4^- \backslash x_2) \\ (T_2^- \backslash x_1) \wedge (T_7 \backslash x_2) \end{pmatrix}$$

$$fh_{\text{hom}\,2}\begin{pmatrix} x_0 \\ x(k-1) \\ x(k) \\ x(k+1) \\ u(k) \end{pmatrix} = \begin{pmatrix} +\infty \\ +\infty \\ [A_1 \otimes x(k-1) \oplus B \otimes u(k)] \\ \wedge f^+(x(k-1), u(k)) \\ +\infty \\ +\infty \end{pmatrix}$$

with $A_1 \otimes x(k) \oplus B \otimes u(k) = \begin{cases} T_9 \otimes x_2(k-1) \oplus T_6 \otimes u_1(k) \\ T_8 \otimes x_1(k-1) \oplus T_7 \otimes u_2(k) \end{cases}$ and

$$f^+(x(k-1), u(k)) = \begin{cases} T_1^+ \otimes u_1(k) \oplus T_2^+ \otimes u_2(k) \\ (T_3^+ \otimes x_1(k-1)) \wedge (T_4^+ \otimes u_1(k)). \\ \wedge(T_5^+ \otimes x_2(k-1)) \end{cases}$$

4.6.3. *Existence*

The only circuit of $fh_{\text{hom},1}$ is the loop connected to x_0 whose weight equals zero. Let us consider a horizon $l = 6$: $X_6 = (x_0, x_1(k_s), x_2(k_s), u_1(k_s), u_2(k_s), ..., x_1(k_s+6), x_2(k_s+6), u_1(k_s+6), u_2(k_s+6))^t$. As there is a path from the above circuit to every vertex except $x_1(k_s+6)$, the relevant components in $\chi(\Phi_{fh_{\text{hom},1}}(X_6))$ are equal to zero. The associated graph of fh_{hom} is far more complicated than $fh_{\text{hom},1}$ and its treatment requires a program specific to dynamic systems. The algorithm in section 4.5 gives a simple approach: as the calculations show the existence of a state and input trajectory for a horizon $[k_s, k_s+6]$, Theorem 4.4 gives $\chi(\Phi_{fh_{\text{hom}}}(X_6)) \geq 0$. We can deduce that $0 \leq \chi(\Phi_{fh_{\text{hom}}}(X_6)) \leq \chi(\Phi_{fh_{\text{hom},1}}(X_6))$. Moreover, there is a path in the associated graph of fh_{hom}, from x_0 to $x_1(k_s+6)$ (the relations are $x_1(k) \leq T_9 \otimes x_2(k-1) \oplus T_6 \otimes u_1$ and $x_2(k-1) \leq z(k-1) \otimes x_0$). Consequently, $\chi(\Phi_{fh_{\text{hom}}}(X_6)) = 0$.

We now give some examples of simple circuits, which illustrate the possible incoherencies in the plant (circuit 1) or, between the plant and its specifications (circuits 2 and 3). When a strictly negative circuit appears, the algorithm of control diverges to infinite values: if condition

$\chi(\Phi_{fh_{\text{hom}}}(X_l)) \geq 0$ is not satisfied, Theorem 4.4 states that there is no finite solution.

Circuit 1

$$\begin{cases} x_2(k) \leq T_5^- \setminus x_2(k+1)) \\ x_2(k+1) \leq (T_5^+ \otimes x_2(k-1)) \end{cases}$$

For instance, an elementary error in the modeling is the consideration of $T_5^- > T_5^+$ with $T_5^- = 10$ and $T_5^+ = 3$. The weight of the circuit is $T_5^+ - T_5^-$. Therefore, the mean weight of the circuit is negative and equal to $-7/2$. The calculation of the spectral vector with an algorithm developed with Scilab confirms this result and shows the incoherence of the process on horizon l=1: $\chi(\Phi_{fh_{\text{hom}}}(X_1)) = (0, -3.5, -3.5, -3.5, -3.5, -3.5, -3.5, -3.5, -3.5)^t$.

Circuit 2

$$\begin{cases} x_1(k) \leq ((T_3^- \oplus T_8) \setminus x_2(k+1)) \\ x_2(k+1) \leq (T_3^+ \otimes x_1(k)) \end{cases}$$

The weight of the circuit is $T_3^+ - (T_3^- \oplus T_8) = 11 - (2 \oplus 3) = 8$.

Now, if $T_8 = 12$, the mean weight of the circuit is negative and equal to -0.5. The calculation of the spectral vector confirms this result and shows that the process cannot follow the specifications on horizon l=1: $\chi(\Phi_{fh_{\text{hom}}}(X_1)) = (0, -0.5, -0.5, -0.5, -0.5, -0.5, -0.5, -0.5, -0.5)^t$.

Circuit 3

$$\begin{cases} x_2(k) \leq T_4^+ \otimes u_1(k) \\ u_1(k) \leq (T_6 \oplus T_1^-) \setminus x_1(k)) \\ x_1(k) \leq (T_3^- \oplus T_8) \setminus x_2(k+1) \\ x_2(k+1) \leq T_5^+ \otimes x_2(k) \end{cases}$$

The weight of the circuit is $T_4^+ - (T_6 \oplus T_1^-) - (T_3^- \oplus T_8) + T_5^+ = 10 - (6) - 3 + 3 = 4$.

Now, if $T_4^+ = 5$, the mean weight of the circuit is negative and equal to -0.25. The calculation of the spectral vector confirms this result and shows that the process cannot follow the specifications.

4.6.4. *Optimal control with specifications*

The desired output z is as follows:

k	k_s	k_s+1	k_s+2	k_s+3	k_s+4	k_s+5	k_s+6
z	30	31	32	33	34	35	35

The resolution of part $fh_{\text{hom},1}$ gives:

k	k_s	k_s+1	k_s+2	k_s+3	k_s+4	k_s+5	k_s+6
x_1	22	22	27	27	32	32	$+\infty$
x_2	20	25	25	30	30	35	35

k	k_s	k_s+1	k_s+2	k_s+3	k_s+4	k_s+5	k_s+6
u_1	16	16	21	21	26	26	32
u_2	19	20	24	25	29	30	34

Now, the complete model fh_{hom} is considered. The following table gives the greatest trajectories of the state and the control for different initial conditions. Let us note that the previous greatest solution gives a finite upper bound of the following solutions that satisfied $V = fh(V)$ (see Proposition 4.2). Equally, the program checks that the following trajectories satisfy the model of the timed event graph and its specifications.

For $x_{k_s} = \begin{pmatrix} +\infty \\ +\infty \end{pmatrix}$

k	k_s	k_s+1	k_s+2	k_s+3	k_s+4	k_s+5	k_s+6
x_1	20	22	25	27	30	32	38
x_2	20	23	25	28	30	33	35

k	k_s	k_s+1	k_s+2	k_s+3	k_s+4	k_s+5	k_s+6
u_1	14	16	19	21	24	26	32
u_2	18	20	23	25	28	30	34

The above trajectories represent the greatest trajectories that cannot be exceeded.

4.6.5. *Initial conditions*

Now, the consistency or admissibility of $x(k_s) = x_{k_s}$ is considered. For $x_{k_s} = \begin{pmatrix} 19 \\ 20 \end{pmatrix}$

k	k_s	k_s+1	k_s+2	k_s+3	k_s+4	k_s+5	k_s+6
x_1	19	22	25	27	30	32	38
x_2	20	22	25	28	30	33	35

Therefore, $x(k_s) = x_{k_s}$ is admissible.

k	k_s	k_s+1	k_s+2	k_s+3	k_s+4	k_s+5	k_s+6
u_1	13	16	19	21	24	26	32
u_2	17	20	23	25	28	30	34

For $x_{k_s} = \begin{pmatrix} 18 \\ 20 \end{pmatrix}$

k	k_s	k_s+1	k_s+2	k_s+3	k_s+4	k_s+5	k_s+6
x_1	18	21	24	27	30	32	38
x_2	19	21	24	27	30	33	35

Therefore, $x(k_s) \neq x_{k_s}$is not admissible.

k	k_s	k_s+1	k_s+2	k_s+3	k_s+4	k_s+5	k_s+6
u_1	12	15	18	21	24	26	32
u_2	16	19	22	25	28	30	34

4.7. Conclusion

In this chapter, we have shown that different time event graphs can be modeled in the form of a new model called a time interval system which is based on (min, max, +) functions. The interval descriptor system can describe the time behavior of many models such as timed event graphs, P-time Petri nets and time stream event graphs. These results lead us to assume that the interval descriptor system will bring a unified model and a new subject to the field of time and timed Petri nets.

Considering the semantic rules "And" and "Weak-And" in the case of a time stream event graph, the control problem is reformulated under a fixed-point form. The spectral theory gives conditions of existence of a solution that satisfies the corresponding equality when the event graph is structurally observable. A proposed algorithm that is pseudo-polynomial makes it possible to determine the greatest state and control when the timed event graph starts from an arbitrary initial condition. The application of the calculated control generates a state trajectory following the specifications and checks the initial condition.

5

Online Aspect of Predictive Control

In Chapter 4, we focused on the trajectory-tracking control of a timed event graph with specifications defined by interval models. In this chapter, we focus on a sliding horizon when the specifications are defined by some P-time event graphs. A crucial point is that too slow a calculation of the control can postpone the application of the control at the calculated dates. Making the most of the specific structures of the systems, the CPU (central processing unit) time of the online procedure is drastically reduced with two techniques: the use of specific algorithms of graph theory instead of generic algorithms and an offline preparation that avoids the useless repetition of the same calculations. As a result, the approach can deal with long horizons and important systems, and the application field of the trajectory-tracking control is expanded.

5.1. Introduction

5.1.1. *Problem*

Naturally, the topic of predictive control has already been considered in different studies [SCH 01] in which a usual step is to transform the (max, +) problem into a linear programming problem in the conventional algebra that allows the application of classical algorithms. The principal advantage of this technique is that it takes into consideration the different classes of models. However, model predictive control is an online approach that needs *efficient algorithms*: indeed, the crucial point is that too slow a calculation of the control can postpone the application of the control at the calculated dates.

The extended linear complementary problem (ELCP) algorithm described in Chapter 3 from the thesis by Bart de Schutter cannot be used for online computations as the CPU time increases exponentially [SCH 01]. The algorithms proposed by Khashiyan and Karmarkar in linear programming are well-known algorithms but they are polynomial *in the weak sense* (the complexities of the ellipsoid algorithm of Khashiyan and the interior point algorithm of Karmarkar are $O(n^4.L)$ and $O(n^{3.5}.L)$, respectively where n is the number of variables and L is the number of bits necessary in the storage of the data [SCH 87] [SAV 01]) and not *in the strong sense* (contrary to many algorithms in graph theory: see Chapter 2 in [GON 84]). The complexity of the simplex is exponential in the worst case even if this algorithm is relatively good on average [PAN 85]. As a result, the application of these *generic* algorithms of linear programming (see the algorithms shown in [SCH 01]) leads to a limitation of the size of the systems considered and the magnitude of the entries. Another difficulty is that these generic algorithms correspond to the optimization stage that needs to start from an admissible solution resulting from another stage. This determination can be included in the program (function simplex "linpro()" of Scilab) or not (function "karmarkar(a,b,c,x0)" of Scilab where x0 is the initial vector). Therefore, the problem is the determination of an admissible solution that is close to the optimal solution.

5.1.2. *Specific characteristics*

In fact, the crucial point is that the structure of the matrices considered above presents *specific characteristics*: the matrices are sparse and contain many rows with two non-null entries (1 and -1) at the most. The matrices are close to the in-going/out-going incidence matrices of the fundamental marking relation (see Chapter 3, "Cycle Time"). Therefore the goal of this chapter is to make the most of these specific structures of the systems and to deduce an approach having a reduced CPU time.

A general answer is to use the (max, +) algebra that allows the application of efficient algorithms of path theory: in general, they are *strongly* polynomial (the running time does not depend on the magnitudes of the parameters). A direct consequence is that these algorithms specific to path algebra surpass the best generic algorithms of linear programming when they are applied to the specific problems: these algorithms can consider large-scale systems. Our tests show that if the simplex is used to calculate the maximal paths between every pair of vertices (function "linpro()"), the approach can work until 60 vertices and needs about 2 min while the (max, +) Kleene star (function "star()") solves the problem in 1 s for 300 vertices. Moreover, the (max, +) algebra allows the writing of elegant expressions contrary to linear programming which can give only numerical results.

Another improvement required is as follows. In fact, the predictive control is made on a sliding horizon: the horizon is moved back slightly at each step and the control is calculated. The idea is to avoid the repetition of the same calculations at each step, which can be costly in terms of time. Before the application of the online control, a *preparation* can contain these calculations allowing a reduction of the complexity of the online procedure.

The chapter is structured as follows: in section 5.2, we consider the control problem without desired output and with a fixed horizon and the determination of the earliest trajectory (problem 1). In section 5.3, the desired output defined on a fixed horizon is introduced and a just-in-time control is made (problem 2). The problem is to determine the greatest input in order to obtain a desired behavior defined by the desired output and the specifications. In sections 5.4 and 5.5, the approach is generalized to a sliding horizon (problem 3), and we consider the CPU time of the online procedure based on a preparation using the Kleene star of a large-scale tri-diagonal matrix. Finally, the structure analysis of this special leads to a more efficient technique of calculation of the Kleene star.

As we know that maximization, minimization and addition operations are denoted by $\oplus$, $\wedge$ and $\otimes$, respectively. A matrix is called row-astic if it has no null row.

5.2. Control without desired output (problem 1)

5.2.1. *Objective*

The problem here in this section is the control of a plant described by a timed event graph when the state and control trajectories are constrained by additional specifications defined by a P-time event graph. The objective is the determination of an admissible (arbitrary) control u on horizon $[k_s+1, k_f]$ such that its application to the timed event graph defined by

$$\begin{cases} x(k+1) = A \otimes x(k) \oplus B \otimes u(k+1) \\ y(k) = C \otimes x(k) \end{cases} \qquad [5.1]$$

satisfies the following conditions:

a) The state trajectory follows the model of the autonomous P-time event graph defined by

$$\begin{pmatrix} x(k) \\ x(k+1) \end{pmatrix} \geq \begin{pmatrix} A^{=} & A^{+} \\ A^{-} & A^{=} \end{pmatrix} \otimes \begin{pmatrix} x(k) \\ x(k+1) \end{pmatrix} \qquad [5.2]$$

for $k \geq k_s$; matrix A^- (respectively, A^+) contains the lower bounds of the temporizations (respectively, the upper bounds with minus sign) associated with each place having a unitary initial marking. We have $A^{=} = A_0^- \oplus A_0^+$, where matrix A_0^- (respectively, A_0^+) is defined as A^- (respectively, A^+) but with with a null initial marking. We can refer to Chapter 2 on consistency for more details.

b) The first state vector of the state trajectory $x(k)$ for $k \geq k_s$ is finite and is a known vector denoted as $\underline{x}(k_s)$. This "non-canonical" initial condition can be the result of a past evolution of a process. As $x(k_s)$ is finite, the trajectories considered in this chapter are finite.

Underlined symbols as in $\underline{x}(k_s)$ correspond to the known data of the problem, and $x(k)$ and $y(k)$ are estimated in the following resolutions.

Therefore, we focus on a control problem without desired output on a fixed horizon. This problem is denoted as problem 1.

The following example shows that the space solution to problem 1 is not an inf-semilattice, that is problem 1 has no (unique) minimal solution with respect to the component-wise order.

5.2.2. *Example 1*

Timed event graph:

$$A = \begin{pmatrix} T_1 & \varepsilon \\ T_2 & T_3 \end{pmatrix}, B = \begin{pmatrix} T_4 & T_5 & \varepsilon \\ \varepsilon & T_6 & T_7 \end{pmatrix} \text{ and } C = \begin{pmatrix} \varepsilon & T_8 \end{pmatrix}.$$

P-time event graph:

$$A^{=} = \varepsilon, A^- = \begin{pmatrix} T_{10} & T_{13} \\ T_{11} & T_{12} \end{pmatrix} \text{ and } A^+ = \begin{pmatrix} -T_{20} & -T_{21} \\ -T_{23} & -T_{22} \end{pmatrix}.$$

Every temporization is finite except $T_{20} = T_{21} = T_{22} = T_{23} = +\infty$.

For $k_f = k_s + 1$, we have $x(k_s + 1) = A \otimes \underline{x}(k_s) \oplus B \otimes u(k + 1) \geq A^- \otimes \underline{x}(k_s)$. For $\underline{x}(k_s) = (0\ \ 0)^t$,

$$\begin{cases} x_1(k_s+1) = T_1 \oplus T_4 \otimes u_1(k_s+1) \oplus T_5 \otimes u_2(k_s+1) \\ \geq T_{10} \oplus T_{13} \\ x_2(k_s+1) = T_2 \oplus T_3 \oplus T_6 \otimes u_2(k_s+1) \oplus T_7 \otimes u_3(k_s+1) \\ \geq T_{11} \oplus T_{12} \end{cases}$$

If $T_1 < T_{10} \oplus T_{13}$ and $T_2 \oplus T_3 < T_{11} \oplus T_{12}$, the system becomes:

$$\begin{cases} T_4 \otimes u_1(k_s+1) \oplus T_5 \otimes u_2(k_s+1) \geq T_{10} \oplus T_{13} \\ T_6 \otimes u_2(k_s+1) \oplus T_7 \otimes u_3(k_s+1) \geq T_{11} \oplus T_{12} \end{cases}$$

This system has no (unique) minimal solution with respect to the component-wise order: if $u_2(k_s + 1) = \varepsilon$, $u_1(k_s + 1)$ and $u_3(k_s + 1)$ must be finite; if $u_1(k_s + 1) = \varepsilon$ and $u_3(k_s + 1) = \varepsilon$, $u_2(k_s + 1)$ must be finite. Remember that $\varepsilon = -\infty$. □

In the following section 5.2.3, we present the relations that describe a trajectory of a timed event graph satisfying the specifications defined by a P-time event graph (constraint a).

5.2.3. *Trajectory description*

From [5.1] and [5.2], we deduce a system that describes the trajectories on horizon $[k_s, k_f]$. Let us introduce the following notations: $X = (x(k_s)^t\ x(k_s+1)^t\ x(k_s+2)^t\ \cdots\ x(k_f-1)^t\ x(k_f)^t)^t$ (t: transposed) and

$$D_h = \begin{pmatrix} A^= & A^+ & \varepsilon & \cdots & \varepsilon & \varepsilon & \varepsilon \\ A \oplus A^- & A^= & A^+ & \cdots & \varepsilon & \varepsilon & \varepsilon \\ \varepsilon & A \oplus A^- & A^= & \cdots & \varepsilon & \varepsilon & \varepsilon \\ \cdots & \cdots & \cdots & \cdots & \cdots & \cdots & \cdots \\ \varepsilon & \varepsilon & \varepsilon & \cdots & A^= & A^+ & \varepsilon \\ \varepsilon & \varepsilon & \varepsilon & \cdots & A \oplus A^- & A^= & A^+ \\ \varepsilon & \varepsilon & \varepsilon & \cdots & \varepsilon & A \oplus A^- & A^= \end{pmatrix}$$

with $h = k_f - k_s$. Matrix D_h presents an original block tri-diagonal structure: this is a square matrix, composed of a lower diagonal (square sub-matrices $A \oplus A^-$), a main diagonal (square sub-matrices $A^=$) and an upper diagonal (square sub-matrices A^+), with all other blocks being zero matrices (ε). Matrix D_h is a $n.(h+1)$ x $n.(h+1)$ matrix where n is the dimension of x.

THEOREM 5.1 (Theorem 2 in [DEC 09]).– *The state trajectories of a timed event graph [5.1] starting from* $\underline{x}(k_s)$ *and following the specifications defined by a P-time event graph [5.2] on horizon* $[k_s, k_f]$ *satisfy the following system equation:*

$$\begin{cases} X \geq D_h \otimes X \\ x(k) \geq B \otimes u(k) \text{ for } k \in [k_s+1, k_f] \\ x(k) \leq A \otimes x(k-1) \oplus B \otimes u(k) \text{ for } k \in [k_s+1, k_f] \\ x(k_s) = \underline{x}(k_s) \end{cases} \quad [5.3]$$

REMARK.– It is important to note that system [5.3] cannot be rewritten in a fixed-point form that can be analyzed by known results.

5.2.4. *Relaxed system*

Equality $x(k) = A \otimes x(k-1) \oplus B \otimes u(k)$ comes from the earliest firing rule. In this section, we determine the conditions so that the determination of the trajectory needs to use only $x(k) \geq A \otimes x(k-1) \oplus B \otimes u(k)$. Therefore, the relation $x(k) \leq A \otimes x(k-1) \oplus B \otimes u(k)$ in [5.3] is disregarded. From system [5.3] which describes problem 1, we deduce the following relaxed system

$$\begin{cases} X \geq D_h \otimes X \\ x(k_s) = \underline{x}(k_s) \end{cases} \quad [5.4]$$

which presents a fixed-point form. We now characterize the sets of trajectories of systems [5.3] and [5.4].

PROPERTY 5.1.– Each trajectory of system [5.3] satisfies [5.4].

PROOF.– Immediate: as system [5.3] contains an additional constraint, any trajectory of this system satisfies the relaxed system [5.4]. □

The extended space defined by [5.4] is now restricted by an additional condition $B \otimes u(k) = x(k)$. Contrary to the initial space (see example 1), the space is now an inf-semilattice allowing the use of the efficient algorithms of graph theory. The following theorem is the starting point of the proposed approach. As we consider below the equality $B \otimes u(k) = x(k)$ with $x(k)$ finite for $k \in [k_s + 1, k_f]$, matrix B is necessarily row-astic.

THEOREM 5.2 (Theorem 4 in [DEC 09]).– *A trajectory X of [5.4] satisfies [5.3] if control $u(k)$ satisfies condition $B \otimes u(k) = x(k)$ for $k \in [k_s + 1, k_f]$.*

Therefore, an (arbitrary) admissible trajectory satisfying problem 1 can be found if we can find a trajectory satisfying the relaxed system [5.4] under the condition of existence of a control such that $B \otimes u(k) = x(k)$ for $k \in [k_s + 1, k_f]$. We can also focus on the earliest state trajectory. Let us now determine the earliest state trajectory X^- with [5.4]. Let $E = (\underline{x}(k_s)^t \quad \varepsilon \quad \cdots \quad \varepsilon)^t$ with $| E |=| X |$. As constraint $x(k_s) = \underline{x}(k_s)$ can be written $x(k_s) \leq \underline{x}(k_s)$ and $\underline{x}(k_s) \leq x(k_s)$, the earliest state trajectory X^- is given by the resolution $X \geq D_h \otimes X \oplus E$ with condition $\underline{x}(k_s) \geq x^-(k_s)$. The application of Kleene star by Theorem 4.75 part 1 in [BAC 92] gives the lowest solution $X^- = (D_h)^* \otimes E$ with condition $\underline{x}(k_s) \geq x^-(k_s)$. Moreover, a control can easily be calculated: the greatest control to $B \otimes u(k) = x(k)$ is obviously $u(k) = B \backslash x(k)$ under the condition $B \otimes (B \backslash x(k)) = x(k)$.

Let us now make a brief analysis of $B \otimes u(k) = x(k)$. The analysis of this condition is out of the scope of this chapter and we can expect a study in this direction in more depth in the future. A first analysis on the condition $B \otimes u(k) = x(k)$ is given in [DEC 09]. In the following property, the considered set of matrices include the set of permutation matrices without being limited to this set.

PROPERTY 5.2.– If each row i of the matrix B contains at least one non-null element $B_{i,j}$, which is unique in column j, we can find a control $u(k)$ such that $B \otimes u(k) = x(k)$ is satisfied for any $x(k)$.

PROOF.– If B is row-astic, each row i contains one non-null entry B_{ij} at least and a row can contain more than one element. Suppose that there is an element $B_{i,j}$, which is unique in column j. As a result, the residuation is $(A \backslash b)_i = \bigwedge_{j=1}^{m} A_{ji} \backslash b_j$, where A is an $m \times n$ matrix, we obtain $u_j = (B \backslash x(k))_j = B_{ij} \backslash x_i(k)$ for row i and equality $B_{ij} \otimes u_j = x_i(k)$ is satisfied for any state. As the other entries $B_{ij'}$ of the row i satisfy $B_{ij'} \otimes u_{j'} \leq x_i(k)$, there is a control such that $B_{i,.} \otimes u(k) = x_i(k)$ for any state. The generalization to all rows of B is immediate. □

5.3. Control with desired output (problem 2)

5.3.1. *Objective*

In problem 1, the objective is to calculate a state and a control trajectory of a timed event graph [5.1] starting from $\underline{x}(k_s)$ and following specifications defined by a P-time event graph [5.2] on horizon $[k_s, k_f]$. In this section, we consider the "just-in-time" objective and focus on the greatest state and control trajectory where the following condition must be satisfied: $y \leq \underline{z}$ (denoted condition c) knowing the trajectory of the desired output $\underline{z}$ on a fixed horizon $[k_s+1, k_f]$ with $h = k_f - k_s \in \mathbb{N}$. This problem is denoted as problem 2.

5.3.2. *Fixed-point form*

Using the previous description of the state and control trajectories [5.3], we rewrite the problem under a general fixed-point formulation $x \leq f(x)$ which allows the resolution of problem 2. Let $X^+ = (x^+(k_s)^t\ x^+(k_s+1)^t\ x^+(k_s+2)^t\ \cdots\ x^+(k_f-1)^t\ x^+(k_f)^t)^t$ be the greatest estimate of state trajectory X.

THEOREM 5.3 (Theorem 3 in [DEC 09]).– *The greatest state and control trajectory of problem 2 is the greatest solution of the following fixed-point inequality system*

$$\begin{cases} X \leq D_h \backslash X \\ u(k) \leq B \backslash x(k) \text{ for } k \in [k_s+1, k_f] \\ x(k) \leq [A \otimes x(k-1) \oplus B \otimes u(k)] \wedge C \backslash \underline{z}(k) \\ \text{for } k \in [k_s+1, k_f] \\ x(k_s) \leq \underline{x}(k_s) \end{cases} \qquad [5.5]$$

with condition $\underline{x}(k_s) \leq x^+(k_s)$.

Note that inequality [5.5] is equivalent to inequality [5.3] if we add the constraint $C \otimes x(k) \leq \underline{z}$. If condition $\underline{x}(k_s) \leq x^+(k_s)$ is satisfied, then $\underline{x}(k_s) = x^+(k_s)$. Therefore, the calculated state trajectory for $k \geq k_s$ is consistent with the past evolution $k \leq k_s$. In other words, the timed event graph can follow the calculated trajectory X^+ after k_s which obeys the specifications defined by the P-time event graph.

System [5.5] leads to a fixed-point formulation whose general form is such that $x \leq f(x)$ where f is a (min, max, +) function and can be defined by the following function:

$f = b, x_1, x_2, \ldots, x_n | f \otimes a | f \wedge f | f \oplus f$, where a, b are arbitrary real numbers ($a, b \in \mathbb{R}$). It is important to note that the concept of extremal solution exists in system [5.5] contrary to system [5.3].

Effective calculation of the greatest control can be made by a classical iterative algorithm of [MCM 92] (pseudo-polynomial), which particularizes the algorithm of Kleene to (min, max, +) expressions, or more complex algorithms [WAL 95], [CHE 05].

5.3.3. *Relaxed system*

The following result shows that assumption $x(k) = B \otimes u(k)$ gives a simplified form $x \leq f(x)$ where f is only a (min, +) function.

THEOREM 5.4.– *The greatest state trajectory X^+ of problem 2 is the greatest solution of the following fixed-point inequality system*

$$\begin{cases} X \leq D_h \backslash X \\ x(k) \leq C \backslash \underline{z}(k) \textit{ for } k \in [k_s + 1, k_f] \\ x(k_s) \leq \underline{x}(k_s) \end{cases} \qquad [5.6]$$

with condition $\underline{x}(k_s) \leq x^+(k_s)$ if control $u(k)$ satisfies condition $B \otimes u(k) = x(k)$ for $k \in [k_s + 1, k_f]$.

PROOF.– Theorem 5.2 in section 5.2.4 shows that a trajectory X of [5.4] satisfies [5.3] if control $u(k)$ satisfies condition $B \otimes u(k) = x(k)$ for $k \in [k_s + 1, k_f]$. This result holds if we add the constraint $C \otimes x(k) \leq \underline{z}$. Thus, we can deduce system [5.6]. □

Let us calculate the greatest state of the control problem. Let $E' = (\underline{x}(k_s)^t \ (C \backslash \underline{z}(k_s + 1))^t \ (C \backslash \underline{z}(k_s + 2))^t \ \cdots \ (C \backslash \underline{z}(k_f))^t)^t$.

The system of inequalities [5.6] becomes

$$\begin{cases} X \leq D_h \backslash X \wedge E' \\ x(k) = u(k) \text{ for } k \in [k_s + 1, k_f] \end{cases}$$

with condition $\underline{x}(k_s) \leq x^+(k_s)$. The application of Theorem 4.73 in [BAC 92] gives the greatest solution $X^+ = D_h^* \backslash E'$ with condition $\underline{x}(k_s) \leq x^+(k_s)$.

Like the calculation of the earliest state trajectory, the resolution of relaxed fixed-point form [5.6] uses the Kleene star that improves the resolution: now, the complexity is not pseudo-polynomial but polynomial in the strong sense.

Up to now, problems 1 and 2 have been considered on a fixed horizon. The aim of the following section is the extension of problem 2 to a control approach on a sliding horizon.

5.4. Control on a sliding horizon (problem 3): online and offline aspects

Let us briefly recall the technique of predictive control on a sliding horizon. We assume that each event date of transition firing is available for the current number of event k: at step $k = k_s$, $\underline{u}(k_s)$ and $\underline{x}(k_s)$ are known. A future control sequence $u(k)$ for $k \in [k_s + 1, k_s + h]$ is calculated and only the first element of the optimal sequence (here $u(k_s + 1)$) is applied to the process. At the next number of event $k_s + 1$, the horizon is shifted: at step $k_s + 1$, the problem is updated with new information $\underline{u}(k_s + 1)$ and $\underline{x}(k_s + 1)$ and a new calculation is performed.

However, we must guarantee the coherence of the state trajectory between each iteration, that is the future trajectory $k \geq k_s + 1$ is the extension of the past trajectory ($k \leq k_s$): more formally, the equality $\underline{x}(k_s) = x^+(k_s)$ must be satisfied otherwise control problem 2 has no solution. If we consider that the models of the timed event graph and the specifications cannot be modified, one possibility is to put the desired output back so that problem 2 presents a solution. In this chapter, we choose to solve problem 2 with the modified desired output trajectory $z^m(k) = \underline{z}(k) \oplus y^-(k)$ for $k \in [k_s + 1, k_s + h]$ where y^- is an admissible trajectory. This procedure yields an optimal control for z^m, which can always be applied to the process. Indeed, if u^- and X^- are the control and the state corresponding to the admissible trajectory y^-, then the control and the state corresponding to z^m are greater.

5.4.1. *CPU time of the online control*

As control approaches using a sliding horizon need efficient algorithms, we consider the complexity of the calculation of the online control. We carried out computation tests on the CPU time of the proposed approach using the max-plus toolbox in Scilab 3.1.1 with an Intel Core2 Duo 2.26 GHz. This time value includes the time of the two algorithms corresponding to the prediction procedure and the control synthesis proposed in this chapter which correspond respectively to the resolution of the relaxed systems [5.4] and [5.6] of problems 1 and 2. However, the CPU time of the control does not include the calculation of the star $(D_h)^*$ which is made in the *offline preparation* that depends on the matrices of the model and the size of the horizon. It needs the memorization of a large matrix ($(h+1).n$ x $(h+1).n$). Therefore, the calculations of the online control are limited only to the multiplication of matrices $(D_h)^*$ and E ($X^- = (D_h)^* \otimes E$) and the left $\otimes$−residuation of E' by D_h^* ($X^+ = D_h^* \backslash E'$) in time $O(q^2)$, where $q = (h+1).n$. In Figure 5.1, we have listed the CPU time needed to compute the control for horizon $h = 50$ and different dimensions n of the state matrices of the event graph until $n = 97$. The matrices of the system are randomly generated ($A^{=} = \varepsilon$, A^{-} and A^{+} are full). Note that for $n = 90$, the offline preparation, which is the calculation of the Kleene star with function star() of Scilab, needs approximately 2.10^3 s, whereas the online procedure needs only 0.28 s (Figure 5.1). Therefore, the CPU time of the online control is drastically reduced.

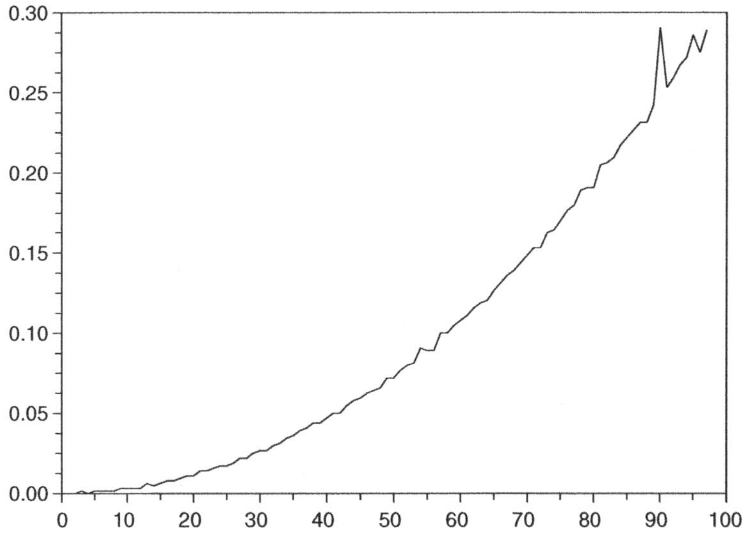

Figure 5.1. *Online control: CPU times of one step for h = 50 and randomly generated matrices until n = 97 transitions. The CPU time limited to 0.3 s shows the efficiency of the procedure*

5.5. Kleene star of the block tri-diagonal matrix and formal expressions of the sub-matrices

The determination of the earliest and greatest trajectories needs the calculation of the star of D_h. As this matrix can be large, the CPU time can be significant (see Figure 5.2) even if it is polynomial in the strong sense.

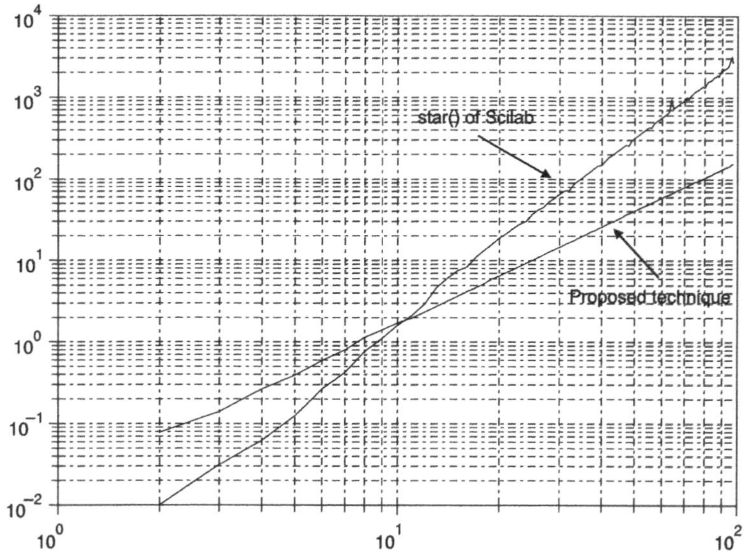

Figure 5.2. *Offline preparation: CPU times of the Kleene star of function star() of Scilab and the proposed technique for* $h = 50$ *and randomly generated matrices*

Making the most of the specific structure of D_h, we present below another technique that gives the formal expression of the sub-matrices of $(D_h)^*$ using the matrices of the system, that is, A, A^-, $A^=$ and A^+. Therefore, we show that $(D_h)^*_{j,j} = [w_j \oplus v_{h-j}]^*$for $j = 0$ to h, where $w_0 = v_0 = A^=$, $w_i = A^= \oplus (A \oplus A^-) \otimes (w_{i-1})^* \otimes A^+$ and $v_i = A^= \oplus A^+ \otimes (v_{i-1})^* (A \oplus A^-) \otimes$ for $i = 1$ to h. Let us present these results.

The star $(D_h)^*$ is the least solution to $Z \geq D_h \otimes Z \oplus I_{(h+1).n}$ and $Z = D_h \otimes Z \oplus I_{(h+1).n}$ where the dimension of matrix Z is $((h+1).n) \times ((h+1).n)$. We note $Z = (Z_{.,0} \quad Z_{.,1} \quad \ldots \quad Z_{.,j} \quad \ldots \quad Z_{.,h})$, where $Z_{.,j}$ for $j = 0$ to h is a matrix composed of n columns j of Z from $1 + j.n$ to $n + j.n$. The system is a set of $h + 1$ systems whose form is

$$Z_{.,j} \geq D_h \otimes Z_{.,j} \oplus N_j \text{ for } j = 0 \text{ to } h\text{, where } N_j = \begin{pmatrix} \varepsilon \\ \dots \\ \varepsilon \\ I_n \\ \varepsilon \\ \dots \\ \varepsilon \end{pmatrix},$$

$$Z_{.,j} = \begin{pmatrix} Z_{0,j} \\ \dots \\ Z_{j-1,j} \\ Z_{j,j} \\ Z_{j+1,j} \\ \dots \\ Z_{h,j} \end{pmatrix}$$ and the identity matrix I_n of N_j and matrix $Z_{j,j}$ have the same position $j+1$ and dimensions.

As the least solution satisfies the equality, the development of the system is as follows.

$$\begin{cases} Z_{0,j} = A^{=} \otimes Z_{0,j} \ \oplus A^{+} \otimes Z_{1,j} \\ Z_{i,j} = (A \oplus A^{-}) \otimes Z_{i-1,j} \oplus A^{=} \otimes Z_{i,j} \ \oplus A^{+} \otimes Z_{i+1,j} \\ \text{for } i = 1 \text{ to } j-1 \\ Z_{j,j} = (A \oplus A^{-}) \otimes Z_{j-1,j} \oplus A^{=} \otimes Z_{j,j} \oplus A^{+} \otimes Z_{j+1,j} \oplus I_n \\ Z_{i,j} = (A \oplus A^{-}) \otimes Z_{i-1,j} \oplus A^{=} \otimes Z_{i,j} \ \oplus A^{+} \otimes Z_{i+1,j} \\ \text{for } i = \ j+1 \text{ to } h-1 \\ Z_{h,j} = (A \oplus A^{-}) \otimes Z_{h-1,j} \oplus A^{=} \otimes Z_{h,j} \end{cases} \quad [5.7]$$

We now consider below the least solution of equations and subsystems of [5.7]. Note that placing partial least solutions in other equations and subsystems is coherent with the objective of getting the least solution of the complete system because all operations involved are isotone.

Let us consider the following subsystem of [5.7] for $i = 0$ to $j - 1$

$$\begin{cases} Z_{0,j} = A^{=} \otimes Z_{0,j} \oplus A^{+} \otimes Z_{1,j} \\ Z_{i,j} = (A \oplus A^{-}) \otimes Z_{i-1,j} \oplus A^{=} \otimes Z_{i,j} \oplus A^{+} \otimes Z_{i+1,j} \\ \text{for } i = 1 \text{ to } j - 1 \end{cases} \qquad [5.8]$$

Let $w_0 = v_0 = A^{=}$, $w_i = A^{=} \oplus (A \oplus A^{-}) \otimes (w_{i-1})^* \otimes A^{+}$ and $v_i = A^{=} \oplus A^{+} \otimes (v_{i-1})^*(A \oplus A^{-})$ for $i = 1$ to h.

PROPERTY 5.3.– The least solution to system [5.8] is $Z_{i,j} = (w_i)^* \otimes A^{+} \otimes Z_{i+1,j}$ for $i = 0$ to $j - 1$.

PROOF.– The proof is based on a recursion. Let $\mathcal{P}(i)$ be defined by $Z_{i,j} = (w_i)^* \otimes A^{+} \otimes Z_{i+1,j}$.

Base case: $\mathcal{P}(0)$

From the first equality of [5.8], we can write $Z_{0,j} = w_0 \otimes Z_{0,j} \oplus A^{+} \otimes Z_{1,j}$ and $Z_{0,j} = (w_0)^* \otimes A^{+} \otimes Z_{1,j}$, where $w_0 = A^{=}$ which proves $\mathcal{P}(0)$.

Case: $\mathcal{P}(i)$ for i from 1 to $j - 1$.

Let us assume $\mathcal{P}(i-1)$: $Z_{i-1,j} = (w_{i-1})^* \otimes A^{+} \otimes Z_{i,j}$. We will prove that $\mathcal{P}(i-1)$ implies $\mathcal{P}(i)$. As $Z_{i,j} = (A \oplus A^{-}) \otimes Z_{i-1,j} \oplus A^{=} \otimes Z_{i,j} \oplus A^{+} \otimes Z_{i+1,j}$, we deduce: $Z_{i,j} = (A \oplus A^{-}) \otimes (w_{i-1})^* \otimes A^{+} \otimes Z_{i,j} \oplus A^{=} \otimes Z_{i,j} \oplus A^{+} \otimes Z_{i+1,j} = [A^{=} \oplus (A \oplus A^{-}) \otimes (w_{i-1})^* \otimes A^{+}] \otimes Z_{i,j} \oplus A^{+} \otimes Z_{i+1,j} = w_i \otimes Z_{i,j} \oplus A^{+} \otimes Z_{i+1,j}$ with $w_i = A^{=} \oplus (A \oplus A^{-}) \otimes (w_{i-1})^* \otimes A^{+}$. Therefore, $\mathcal{P}(i)$ is deduced from $\mathcal{P}(i-1)$. Moreover, as $\mathcal{P}(0)$ is true, $\mathcal{P}(i)$ has been proved for i from 1 to $j - 1$: the recursion is finished. □

Let us consider the following subsystem of [5.7] for $i = j + 1$ to h

$$\begin{cases} Z_{i,j} = (A \oplus A^{-}) \otimes Z_{i-1,j} \oplus A^{=} \otimes Z_{i,j} \oplus A^{+} \otimes Z_{i+1,j} \\ \text{for } i = j + 1 \text{ to } h - 1 \\ Z_{h,j} = (A \oplus A^{-}) \otimes Z_{h-1,j} \oplus A^{=} \otimes Z_{h,j} \end{cases} \qquad [5.9]$$

PROPERTY 5.4.– The least solution to system [5.9] is $Z_{i,j} = (v_{h-i})^* \otimes (A \oplus A^-) \otimes Z_{i-1,j}$ for $i = j+1$ to h.

PROOF.– The proof is based on a recursion as above.

$$\mathcal{P}(i) : Z_{i,j} = (v_{h-i})^* \otimes (A \oplus A^-) \otimes Z_{i-1,j}.$$

Base case: $\mathcal{P}(h)$

From the last equality of [5.9], we can write $Z_{h,j} = (A \oplus A^-) \otimes Z_{h-1,j} \oplus v_0 \otimes Z_{h,j}$ and $Z_{h,j} = (v_0)^* \otimes (A \oplus A^-) \otimes Z_{h-1,j}$, where $v_0 = A^=$ which proves $\mathcal{P}(h)$.

Case: $\mathcal{P}(i)$ for i from $h-1$ to $j+1$.

Let us assume $\mathcal{P}(i+1)$: $Z_{i+1,j} = (v_{h-i-1})^* \otimes (A \oplus A^-) \otimes Z_{i,j}$. We will prove that $\mathcal{P}(i+1)$ implies $\mathcal{P}(i)$. As $Z_{i,j} = (A \oplus A^-) \otimes Z_{i-1,j} \oplus A^= \otimes Z_{i,j} \oplus A^+ \otimes Z_{i+1,j}$, we deduce: $Z_{i,j} = (A \oplus A^-) \otimes Z_{i-1,j} \oplus A^= \otimes Z_{i,j} \oplus A^+ \otimes (v_{h-i-1})^* \otimes (A \oplus A^-) \otimes Z_{i,j} = (A \oplus A^-) \otimes Z_{i-1,j} \oplus [A^= \oplus A^+ \otimes (v_{h-i-1})^* \otimes (A \oplus A^-)] \otimes Z_{i,j} = (A \oplus A^-) \otimes Z_{i-1,j} \oplus v_{h-i} \otimes Z_{i,j}$ with $v_{h-i} = A^= \oplus A^+ \otimes (v_{h-i-1})^* \otimes (A \oplus A^-)$. Therefore, $\mathcal{P}(i)$ is deduced from $\mathcal{P}(i+1)$. Moreover, as $\mathcal{P}(h)$ is true, $\mathcal{P}(i)$ has been proved for i from h to $j+1$: the recursion is finished. □

Up to now, the two previous properties cannot calculate the sub-matrices $(D_h)^*_{i,j}$. The following theorem solves this problem by giving the starting point for each column j.

THEOREM 5.5.–

$$\begin{cases} (D_h)^*_{i,j} = (w_i)^* \otimes A^+ \otimes (D_h)^*_{i+1,j} \textit{ for } i = 0 \textit{ to } j-1 \\ (D_h)^*_{i,j} = (v_{h-i})^* \otimes (A \oplus A^-) \otimes (D_h) \textit{ for } i = h \textit{ to } j+1 \end{cases}$$

with $(D_h)^*_{j,j} = [w_j \oplus v_{h-j}]^*$ for $j = 0$ to h.

PROOF.– Properties 5.3 and 5.4 give $Z_{j-1,j} = (w_{j-1})^* \otimes A^+ \otimes Z_{j,j}$ and $Z_{j+1,j} = (v_{h-j-1})^* \otimes (A \oplus A^-) \otimes Z_{j,j}$ which are two specific cases.

Therefore, $Z_{j,j} = (A \oplus A^-) \otimes Z_{j-1,j} \oplus A^= \otimes Z_{j,j} \oplus A^+ \otimes Z_{j+1,j} \oplus I_n = [(A \oplus A^-) \otimes (w_{j-1})^* \otimes A^+ \oplus A^= \oplus A^+ \otimes (v_{h-j-1})^* \otimes (A \oplus A^-)] \otimes Z_{j,j} \oplus I_n$

Finally, $Z_{j,j} = [(A \oplus A^-) \otimes (w_{j-1})^* \otimes A^+ \oplus A^= \oplus A^+ \otimes (v_{h-j-1})^* \otimes (A \oplus A^-)]^* = [A^= \oplus (A \oplus A^-) \otimes (w_{j-1})^* \otimes A^+ \oplus A^= \oplus A^+ \otimes (v_{h-j-1})^* \otimes (A \oplus A^-)]^* = [w_j \oplus v_{h-j}]^*$ which is the sub-matrix (j, j) of $(D_h)^*$.

We can also deduce every matrix of column j of $(D_h)^*$ from properties 5.3 and 5.4:

$$(D_h)^*_{i,j} = (w_i)^* \otimes A^+ \otimes (D_h)^*_{i+1,j} \text{ for } i = 0 \text{ to } j - 1 \text{ and}$$

$$(D_h)^*_{i,j} = (v_{h-i})^* \otimes (A \oplus A^-) \otimes (D_h)^*_{i-1,j} \text{ for } i = h \text{ to } j + 1. \quad \square$$

We present below the developed form of $(D_h)^*_{.,j}$, which gives another presentation of the results.

$$(D_h)^*_{.,j} = \begin{pmatrix} \dots \\ (D_h)^*_{j-2,j} \\ (D_h)^*_{j-1,j} \\ (D_h)^*_{j,j} \\ (D_h)^*_{j+1,j} \\ (D_h)^*_{j+2,j} \\ \dots \end{pmatrix} =$$

$$\begin{pmatrix} \dots \\ (w_{j-2})^* \otimes A^+ \otimes (w_{j-1})^* \otimes A^+ \otimes [w_j \oplus v_{h-j}]^* \\ (w_{j-1})^* \otimes A^+ \otimes [w_j \oplus v_{h-j}]^* \\ [w_j \oplus v_{h-j}]^* \\ (v_{h-j-1})^* \otimes (A \oplus A^-) \otimes [w_j \oplus v_{h-j}]^* \\ (v_{h-j-2})^* \otimes (A \oplus A^-) \otimes (v_{h-j-1})^* \otimes (A \oplus A^-) \\ \otimes [w_j \oplus v_{h-j}]^* \\ \dots \end{pmatrix}$$

REMARK.– $(D_h)^*_{0,0} = (v_h)^*$ and $(D_h)^*_{h,h} = (w_h)^*$.

CPU time

We now compare our method with the direct application of function star() of the max-plus toolbox for Scilab 3.1.1 to matrix D_h. Our approach also uses function star() of this max-plus toolbox but for calculating v_i^* and w_i^*. It is known that the complexity of the star is polynomial in the strong sense $O(n^a)$ with $a = 3$ for the algorithm of Karp corresponding to $O(h^a.n^a)$ in our problem. The proposed approach makes the most of the specific structure of D_h as its complexity is $O(h.n^a)$.

The matrices of the system are randomly generated ($A^= = \varepsilon$, A^- and A^+ are full), and the program verifies that the two calculated $(D_h)^*$ are identical. Computation tests on CPU time are performed with an Intel Core2 Duo 2.26 GHz. We take $h = 50$ and the dimension n of the matrices varies from 2 to 97; therefore, the considered maximal size of D_h is (4947×4947). Figure 5.2 shows a significant improvement in the CPU time: the time gain is about 16 min for $n = 80$.

5.6. Conclusion

Contrary to known model predictive control approaches, the numerical online control is based on a formal preparation and an important part of the calculations are performed offline during preparation. In this chapter, we show that this separation between the online preparation and the offline control, where an important part of the calculations is performed, allows the consideration of important systems (up to 97 transitions) for long horizons ($h = 50$) corresponding to the handling of a large-scale matrix (4947×4947). In fact, the calculation of the Kleene star of a tri-diagonal matrix can be made once as *it does not depend on the desired output trajectory*: it depends on the models and the size of the given horizon only. In that case, the CPU time of the initial approach (which needs approximately 2.10^3 s) is replaced by a new online procedure that needs only 0.28 s. Therefore, the application field of the model predictive control is expanded: the approach can be applied to long horizons and systems with a reduced CPU time.

A preparation is also made in closed-loop approaches where the computation of a linear feedback and of the maximal set of the initial states is performed offline [KAT 07] and the control is calculated online. Remember that approaches based on a feedback defined by a Petri net are limited by the condition that the temporization and the initial marking of each added place are non-negative. Our practical problem is different as the initial condition is uncontrollable, which is the usual assumption of the predictive control approaches. One difficulty is the stabilization of the sequence of semi-modules. R. Katz have given some sufficient conditions on the specifications (sections III and IV in [KAT 07]). These difficulties are avoided in this chapter as we define a subspace which is an inf-semilattice.

The offline preparation of the approach described in the chapter uses the Kleene star of a special block tri-diagonal matrix denoted as D_h, which is often encountered in numerical solutions of engineering problems (e.g. computational fluid dynamics, finite element method). Making the most of the specific structure of D_h, we give the formal expression of the sub-matrices of the star of these matrices using the matrices of the system, that is, A, A^-, $A^=$ and A^+. We show that this technique improves the CPU time of the offline preparation with respect to the direct application of the function star() of Scilab.

Therefore, we have described an approach with a preparation which is close to the closed-loop technique where the corrector is calculated offline. In a future study, we will show that a predictive control using a space corrector can also be made.

Bibliography

[BAC 92] BACCELLI F., COHEN G., OLSDER G.J., QUADRAT J.P., *Synchronization and Linearity: An Algebra for Discrete Event Systems*, Wiley, New York, 1992, available at http://maxplus.org.

[BER 91] BERTHOMIEU B., DIAZ M., "Modeling and verification of time dependent systems using time Petri nets", *IEEE Transactions on Software Engineering*, vol. 17, no. 3, pp. 259–273, 1991.

[BER 07] BERTHOMIEU B., LIME D., ROUX O.H., VERNADAT F., "Reachability problems and abstract state spaces for time Petri nets with stopwatches", *Journal of Discrete Event Dynamic Systems*, vol. 17, pp. 133–158, 2007.

[BUR 91] BURNS S.M., Performance analysis and optimization of asychronous circuits, PhD, Institute of Technology, Pasadena, California, CA, 1991.

[BUT 84] BUTKOVIC P., HEGEDÜS G., "An elimination method for finding all solutions of the system of linear equations over an extremal algebra", *Ekonomicko-Matematicky Obzor*, vol. 20, no. 2, pp. 203–215, 1984.

[CAM 92] CAMPOS J., CHIOLA G., COLOM J.M., SILVA M., "Properties and performance bounds for timed marked graphs", *IEEE Transactions on Circuits and Systems*, vol. 39, no. 5, pp. 386–401, May 1992.

[CHE 02] CHENG Y., ZHENG D.-Z., "On the cycle time of non-autonomous min-max systems", *Proceedings of the Sixth International Workshop WODES on Discrete Event Systems, IEEE Conference Publications*, pp. 411–412, 2002.

[CHE 05] CHENG Y., ZHENG D.-Z., "Min-max inequalities and the timing verification problem with max and linear constraints, discrete event dynamic systems: theory and applications", *Springer Science and Business Media*, vol. 15, pp. 119–143, 2005.

[COC 99] COCHET-TERRASSON J., GAUBERT S., GUNAWARDENA J., "A constructive fixed point theorem for min-max functions", *Dynamics and Stability of Systems*, vol. 14, no. 4, pp. 407–433, 1999.

[COC 01] COCHET-TERRASSON J., Algorithmes d'itération sur les politiques pour les applications monotones contractantes, PhD Thesis, Ecole Nationale Supérieure des Mines de Paris, 2001.

[COC 93] COHEN G., GAUBERT S., QUADRAT J.-P., "From first to second-order theory of linear discrete event systems", *1st SIAC World Congress*, Sydney, Australia, 18–23 July 1993.

[COH 94] COHEN E., MEGIDDO N., "Improved algorithms for linear inequalities with two variables per inequality", *SIAM Journal on Computing*, vol. 23, no. 6, pp. 1313–1347, December 1994.

[COL 03] COLLART-DUTILLEUL S., YIM P., "Time window specification and validation with Petri nets", *ETFA, IEEE Conference on Emerging Technologies and Factory Automation*, Lisbon, Portugal, pp. 232–237, 16–19 September 2003.

[COT 72] COTTLE R.W., VEINOTT A.F., "Polyhedral sets having a least element", *Mathematical Programming*, vol. 3, pp. 238–249, 1972.

[COU 96] COURTIAT J.P., DIAZ M., DE OLIVIERA R.C., SÉNAC P., "Formal models for the description of timed behaviors of multimedia and hyper media distributed systems", *Computer Communications*, vol. 19, pp. 1134–1150, 1996.

[CUL 94] CULIOLI J.-C., "Introduction à l'optimisation", *Ellipses*, p. 316, 1994.

[DEC 05a] DECLERCK P., DIDI ALAOUI M.K., "Optimal control synthesis in interval descriptor systems: application to time stream event graphs", *Proceedings of the 16th IFAC World Congress*, vol. 16, no. 1, Elsevier, Prague, Czech Republic, 4–8 July, 2005, available at http://www.istia.univ-angers.fr/~declerck/.

[DEC 05b] DECLERCK P., DIDI ALAOUI M.K., "Extremal trajectories in P-time Event Graphs: application to control synthesis with specifications", *Proceedings of the 44th IEEE Conference on Decision and Control and European Control Conference ECC, CDC-ECC'05*, Seville, Spain, pp. 7621–7626, 12–15 December, 2005, available at http://www.istia.univ-angers.fr/~declerck.

[DEC 06] DECLERCK P., DIDI ALAOUI M.K., "Liveness and acceptable trajectories in P-time event graphs", *WODES 2006*, Ann Arbor, MI, 10–12 July 2006, available at http://www.istia.univ-angers.fr/~declerck/.

[DEC 07a] DECLERCK P., "Detection of changes by observer in timed event graphs and time stream event graphs", *IFAC DCDS'07*, Cachan, France, 13–15 June 2007, available at http://www.istia. univ-angers.fr/~declerck.

[DEC 07b] DECLERCK P., GUEZZI A., BOIMOND J.-L., "Cycle time of P-time event graphs", *The 4th International Conference on Informatics in Control, Automation and Robotics (ICINCO 2007 Special Session: From Planning to Control of Manufacturing Systems*, Angers, France, 9–12 May 2007, available at http://www.istia.univ-angers.fr/~declerck/publications.html.

[DEC 08] DECLERCK P., GUEZZI A., GROS C., "Temps de cycle des Graphes d'Evénements Temporisés et P-temporels", *Conférence Internationale Francophone d'Automatique, CIFA'2008*, Bucharest, Romania, 3–9 September 2008.

[DEC 09] DECLERCK P., GUEZZI A., "Trajectory tracking control of a timed event graph with specifications defined by a P-time event graph", *POSTA 09 Positive Systems: Theory and Applications*, Valence, Spain, 2–4 September 2009.

[DEC 10a] DECLERCK P., "Trajectory tracking control of a timed event graph with specifications defined by a P-time event graph: on-line control and off-line preparation", *WODES 2010*, Berlin, Germany, 30 August–1 September 2010.

[DEC 10b] DECLERCK P., DIDI ALAOUI M.K., "Optimal control synthesis of timed event graphs with interval model specifications", *IEEE Transactions on Automatic Control*, vol. 55, no. 2, pp. 518–523, February 2010.

[DEC 11] DECLERCK P., "From extremal trajectories to consistency in P-time Event Graphs", *IEEE Transactions on Automatic Control*, vol. 56, no. 2, pp. 463–467, February 2011.

[DEF 08] DEFOSSEZ F., COLLART-DUTILLEUL S., BON P., "Temporal requirements checking in a safety analysis of railway systems, FORMS/FORMAT 2008", *Symposium on Formal Methods for Automation and Safety in Railway and Automotive Systems*, TU Braunschweig and Budapest University of Technology and Economics, October 2008.

[DEN 05] DENAT J.P., COLLART-DUTILLEUL S., CHETOUANE F., "Cycle time calculation for electroplating lines steady state operation using P-time strongly connected event graphs", *Proceedings of IMACS*, Paris, France, 11–15 July 2005.

[DIA 97] DIAZ M., OWEZARSKI P., "From multimedia models to multimedia transport protocols", *Computer Networks and ISDN Systems*, vol. 29, pp. 745–758, 1997.

[DIA 01] DIAZ M., *Les Réseaux De Petri: Modèles Fondamentaux*, Hermès, Paris, France, 2001.

[DID 05] DIDI ALAOUI M.K., Etude et supervision des graphes d'événements temporisés et temporels: vivacité, estimation et commande, PhD Thesis, University of Angers, France, 2005, available at http://www.istia.univ-angers.fr/~declerck/resume-these-khalid.htm.

[GAB 76] GABBAY H., "A note on polyhedral sets having a least element", *Mathematical Programming*, vol. 11, pp. 94–96, 1976.

[GAU 98] GAUBERT S., GUNAWARDENA J., "The duality theorem for min-max functions", *Computer Assisted Radiology and Surgery*, vol. 326, no. I, pp. 43–48, 1998.

[GIU 00] GIUA A., PICCALUGA A., SEATZU C., "Optimal token allocation in timed cyclic event-graphs", *Proceedings of the 4th Workshop on Discrete Event Systems*, IEEE Conference Publications, pp. 209–218, August 2000.

[GON 84] GONDRAN M., MINOUX M., *Graphs and Algorithms*, John Wiley & Sons, March 1984.

[GON 02] GONDRAN M., MINOUX M., *Graphes, dioïdes et semi-anneaux*, Tec et Doc, 2002.

[GUE 09] GUEZZI A., DECLERCK P., BOIMOND J.-L., "Commande de graphes d'événements temporisés sur un horizon glissant", *Journal Européen des Systèmes Automatisés (JESA)*, vol. 43, no. 7–9, pp. 1097–1111, 2009.

[GUE 10] GUEZZI A., Modélisation, analyse de performance et commande des systèmes à événements discrets, PhD Thesis, University of Angers, France, 2010, available at http://www.istia.univangers.fr/.

[HAS 99] HASHTRUDI ZAD S., KWONG R.H., WONHAM W.M., "Supremum operators and computation of supremal elements in system theory", *SIAM Journal on Control and Optimization*, vol. 37, no. 3, pp. 695–709, 1999.

[HEI 06] HEIDERGOTT B., OLSDER G.J., VAN DER WOUDE L., *Max Plus at Work*, Princeton University Press, 2006.

[KHA 96] KHANSA W., DENAT J.-P., COLLART DUTILLEUL S., "P-time Petri nets for manufacturing systems", *WODES'96*, Edinburgh, UK, 1996.

[KHA 97] KHANSA W., Réseaux de Petri P-temporels. Contribution à l'étude des systèmes à Evénements discrets, PhD Thesis, University of Savoie, 7 March 1997.

[KUM 95] Kumar R., Garg V.K., "Extremal solutions of inequations over lattices with applications to supervisory control", *Theoretical Computer Science*, vol. 148, pp. 67–92, 1995.

[KAT 07] KATZ R.D., "Max-plus (A,B)-invariant spaces and control of timed discrete-event systems", *IEEE Transactions on Automatic Control*, vol. 52, no. 2, pp. 229–241, February 2007.

[LIB 96] LIBEAUT L., Sur l'utilisation des dioïdes pour la commande des systèmes à événements discrets, PhD Thesis, University of Nantes and École Centrale de Nantes, France, 1996.

[MAG 84] MAGOTT J., "Performance evaluation of concurrent systems using Petri nets", *Information Processing Letters*, vol. 18, pp. 7–13, 1984.

[MAG 07] MAGNIN M., Réseaux de Petri à chronomètres: temps dense et temps discret, PhD Thesis, Ecole Centrale de Nantes, France, December 2007.

[MCM 92] MCMILLAN K., DILL D., "Algorithms for interface timing verification", *Proceedings of the IEEE, International Conference on Computer Design: VLSI in Computers and Processors*, Los Alamitos, CA, IEEE Comput. Soc. Press, pp. 48–51, 11–14 October 1992.

[MUD 05] MUDER J., "Comments on "timed Petri nets in modeling and analysis of cluster tools", *IEEE Transactions on Automation Science and Engineering*, vol. 2, no. 1, pp. 92–93, January 2005.

[MUR 89] MURATA T., "Petri nets: properties", *Analysis and Applications, Proceedings of the IEEE*, vol. 77, no. 4, pp. 541–580, 1989.

[PAN 85] PAN V., "On the complexity of a pivot step of the revised simplex algorithm", *Computer and Mathematics with Applications*, vol. 11, no. 11, pp. 1127–1140, 1985.

[RAM 80] RAMAMOORTHY C.V., GARY HO S., "Performance evaluation of asynchronous concurrent systems using Petri nets", *IEEE Transactions on Software Engineering*, vol. SE-6, no. 5, pp. 440–449, September 1980.

[SAV 01] SAVARD G., Introduction aux méthodes de points intérieurs, Extract from notes, February 2001.

[SCH 87] SCHRIJVER A., *Theory of Linear and Integer Programming*, John Wiley & Sons, 1987.

[SCH 01] DE SCHUTTER B., VAN DEN BOOM T., "Model predictive control for max-plus-linear discrete event systems", *Automatica*, vol. 37, no. 7, pp. 1049–1056, 2001.

[VAN 05] VAN DER WOUDE J., OLSDER G.J., "On (min,max,+)-inequalities", *Contemporary Mathematics*, vol. 377, pp. 353–361, 2005.

[VEI 68] VEINOTT A., "Extreme points of Leontief substitution systems", *Linear Algebra and Its Applications*, vol. 1, pp. 181–194, 1968.

[WAL 95] WALKUP E., Optimization of linear max-plus systems with application to timing analysis, PhD Thesis, University of Washington, 1995.

[ZEI 86] ZEIDLER E., *Nonlinear Functional Analysis and Its Applications I. Fixed-Point Theorems*, Springer-Verlag, New York, NY, 1986.

List of Symbols

a, b	coefficients of affine function
c	circuit
A, B, C	matrices
$\mathbb{A}$	matrix
Co	set of pairs
Cr	criterion
D_h	matrix on horizon h
e	neutral element of operation $\otimes$ (addition): $e = 0$
E	vector
E'	vector
EG	Event Graph
f	function
fh	function
f_{hom}	homogeneous function
$F(n, n)$	set of functions
$G(A)$	associated graph of A
$G(A, B, C)$	associated graph of A, B and C
h	horizon
H	matrix

I_j	set of upstream arcs of transtion j
k	number of events
L	environ function
Lb	lower bound
M	marking
M_0	initial marking
m	maximal initial marking
$(M_0)_l$, m_l	initial marking of place l
N	used in the calculation of the Kleene star of the block tridiagonal matrix
O	matrix
P	set of places
p	place p
pa	path
Q	matrix
R	set of functions
S	functions
ST	subset of transitions
Sv	set of vertices
Sx	set
$\mathcal{SX}$	complete lattice
$\mathcal{SY}$	set of fixed points
t	transpose
u	control
un	vector $un = (1, 1, ..., 1)^t$
Ub	upper bound
V	vector
W	incidence matrix
x	state vector or arbitrary solution of a system
y	output
z	desired output
Z	intermediate data
α	weight of criterion Cr
β	intermediate vector

γ	row-vector (generalized P-invariant)
δ	data in algorithm
ε	neutral element of operation $\oplus$ (maximum): $\varepsilon = -\infty$
η	ultimately affine regime
θ	fixed point solution
λ	cycle time
μ	$\max\{\mid c\mid_w / \mid c\mid_l \mid i$
υ	ultimately affine regime
ρ	data in algorithm
$\chi(f)$	spectral vector
Δ	shift in the homogeneity property
$\Phi_f(.)$	function
$\mathbb{N}$	set of natural numbers
$\mathbb{R}$	set of real numbers
$\mid P \mid$	cardinal number of set P
$\mid pa \mid_w$	sum of the weights on the edges in the path pa
$\mid pa \mid_l$	number of edges in the path pa
${}^\bullet p$	set of input elements of p
$p^\bullet$	set of output elements of p
$\oplus$	operation maximum
$\otimes$	operation addition
$\wedge$	minimum
$\backslash$	residuation of $\otimes$
A^*	Kleene star of matrix A
A^t	transpose of matrix A
$A_{i,j}$	element (i, j) of matrix A

Index

G, H

I

J, K

L, M

N, O

P

R

S

T

U, W, Z

Printed and bound by CPI Group (UK) Ltd, Croydon, CR0 4YY

13/09/2026

14961075-0008